JOANNE MARCOTTE

Inconvenient Doubts

Climate Change Apocalypse: Really?

SELF-PUBLISHED BY JOANNE MARCOTTE

FROM THE SAME AUTHOR
Pour en finir avec le Gouvernemaman, Éditions Francine Breton Inc., 2011
https://jomarcotte.wordpress.com/

Legal deposit: 4th trimester 2023
Bibliothèque et Archives nationales du Québec, 2023
Library and Archives Canada, 2023
ISBN 978-2-9822039-2-1 (eng.)
ISBN 978-2-9822039-0-7 (original french)

Proofreading: Lara Chauvin
Cover page illustration: Christian Daigle (Fleg)
Cover design and layout: Joanne Marcotte
Back cover photo: Julie Gilbert

Self-published by Joanne Marcotte
Quebec, 2023
https://jomarcotte.wordpress.com/

Printed by Amazon – KDP | direct publishing

To my primary and secondary school teachers,
Thank you.
You were the best.

To all those who have doubts
and who tirelessly seek to
understand our world.

An invitation

This book is an invitation to explore the state of climate science. It offers a counterweight to the climate change doxa* which claims that we are witnessing through our own fault something irremediable: the imminent planetary and humanitarian catastrophe for which mankind is entirely responsible.

It is aimed at people who are still curious and open to debate. Who understand that science is a process of observation and discovery, not a religion. But also, and especially at young people suffering from eco-anxiety and their parents. To you, I say, don't worry so much. It isn't quite what you've been told.

It invites you to meet people who have allowed me, through their work, to open the black box of climate change. Scientists, humanists and economists sincerely and genuinely concerned about the environment and the human condition. Optimistic and confident people who call on us to tone it down, reject apocalyptic alarmism and recognize that we are privileged to live in an era where human progress has made spectacular leaps in such a short space of time.

Never have we heard so much about climate science while censoring those who question it. Still, with this story, I invite you to listen to the little inner voice we all have, which authorizes doubt and claims the freedom to discuss the plausibility of the official narrative on climate change.

---

* Doxa – A set of common opinions considered self-evident.

Introduction

In English, when you say that something is settled, it *is* settled. Settled with a capital 'S'. We understand each other. There is nothing more to discuss. There is a consensus. If there are still those who want to debate, they are excluded from the mainstream media and invited to take refuge on social media platforms.

Consensus of course… because who has not heard that 97% of scientists support the idea of global warming and that the root cause is anthropogenic. In other words, global warming is caused by humans.

Human activity is even *the main cause* given our insatiable thirst for CO_2-generating fossil fuels they say. At least, that is what laymen like me are generously fed by the mainstream media, and what young people are taught from their first years of primary school. The planet is burning, the planet is crying, the planet is suffering from our presence on earth.

But that 97% has always intrigued me. The climate system is extraordinarily complex. We are just tiny ants in an infinitely large universe and on a planet whose climate is influenced by a multitude of natural phenomena that we barely understand. Climate Science with a capital 'S' is in its infancy, the measurement instruments are still imperfect, and climate models have little or no success in reproducing the past. How can we rely on them to predict a climate apocalypse? How can we arrive at such a high level of certainty? Is doubt, the process by which science is refined still allowed?

In December 2015, I wrote a couple of blog posts[1] about how annoyed I get when those who doubt are forbidden to speak out, *Changements climatiques - Criminaliser le doute. Vraiment?*[2] and *COP21 : Le fantasme d'un 'changement de civilisation' par une économie décarbonisée.*[3] Let me remind you: this was in 2015, the year of the Paris Agreement. The year of the COP21, the 'Conference of the Parties', this extraordinary planetary gathering where several countries committed to reducing greenhouse gas emissions, more specifically CO_2 emitted by fossil fuels.

After eight years, my texts haven't aged a day. But if certain 'authority figures' and current media coverage are anything to go by, it seems that the minutes are counting down more than ever before humanity's extinction.

In November 2022, at COP27, the Montreal daily *La Presse*, like many other media, reported on a statement made by the UN Secretary-General Antonio Guterres: "Inaction in the face of the accelerating climate crisis is akin to collective suicide".[4] The CBC referred to a "Highway to Climate Hell".[5] Nothing less.

Alarmist headlines like these have abounded for decades now, without counterweight or nuance. Yet more and more concerned scientists, environmentalists and ecologists are now denouncing the media coverage that surrounds the major international COP fairs, and the publications of the United Nations Intergovernmental Panel on Climate Change (IPCC).

What follows is what I take away from the writings of these authors dedicated to the issue of climate change, the environment and biodiversity.

Is climate science really settled? Not really, says Steven E. Koonin, former Undersecretary for science in the US Department of Energy under President Obama, and author of *Unsettled.*[6] "Still too many unknowns," he believes.

Is there cause for alarm? No, writes Bjorn Lomborg, in his book *False Alarm.*[7] In his view, not only does the catastrophist tone commit astronomical sums that could be redirected to more pressing needs, but it also harms the poorer populations while failing to solve the problem. If you really need to invest large sums of money, let's take the time to do cost/benefit analyses and build smarter solutions, he says.

Then there's early environmentalist Michael Shellenberger, whose *Apocalypse Never*[8] debunks the false claims of climate disinformation propagators. Instead, he proposes a truly humanistic approach that would serve the planet's entire population and its biological diversity.

Finally, I must admit that along the way, I got excited about the new book by climate scientist and specialist in climate uncertainty and risk Judith Curry, *Climate Uncertainty and Risk: Rethinking our Response*[9]. After listening to an interview in which she offered her views on the current state of the debate and on the IPCC's climate models, I decided to read on.

I have read these books. While there are passages that have enlightened me, others were deeply shocking. The level of misinformation by the authorities and the media these authors denounce is troubling. Yet, in my opinion, none of them could be identified as a denier. They all recognize global warming and believe to some extent that CO_2 emissions are a contributing factor. Having said that, all of them suggest toning it down. Science is not a belief system; it is a process of seeking an understanding of the world.

I listened to their interviews with popular podcasters such as Jordan Peterson and Michael Shermer. People we unfortunately do not see in our traditional media, certainly not in Quebec's francophone media. You see, when fear is exploited, clickbait pays off too well politically and in the media.

Here, as elsewhere, some believe that the *climatic doxa* has become the new apocalyptic secular religion. Anyone who does not blindly believe in the official narrative, in the need to atone for the sins of our fossil-fuel-hungry Western civilization, or that profound *'transformational change' is urgent,* is suddenly discredited. As people who doubted sanitary policies were identified as conspiracists, those who doubt that immediate action is required are seen as climate deniers.

There is however good news. These authors offer a much-needed counterweight to the debate. I chose them because they are credible. I also found that they are excellent teachers.

Over time, more and more scientists speak out and try to correct the alarmist trajectory of widely conveyed, if not imposed, opinions. Many are sincerely concerned when a culture of catastrophism and guilt-induced narratives are introduced in our schools. When doom and gloom replace a hopeful future in the mind of youngsters.

Indeed, for several years now, enormous efforts and huge sums are spent to reduce greenhouse gases (GHG), and more specifically, CO_2 emissions. The official narrative goes like this:

- In recent decades, we have seen global warming compared to pre-industrial times (since 1850).
- We have also seen an increase in greenhouse gases (GHG), particularly CO_2 in the atmosphere. This increase is attributed to burning fossil fuels.
- We believe global warming is dangerous, increases the severity of extreme weather events and causes natural disasters.

Therefore, 1) global warming and extreme climate events are due to human activity and 2) we must reduce the use of fossil fuels and opt for the decarbonization of the global economy.

The mission: *urgently and profoundly transform the current engines of prosperity*. Rethink, shake things up and review our way of life. Western civilization is put to the bench, despised and morally reprehensible.

But is there *really* a link between CO_2 emissions and global warming? And to carry out a revolution of such magnitude and in such a radical way (some refer to it as a 'Great Reset'), should we not be damn sure of what we're doing? *How exactly does human activity factor in global warming*? And are there not alternative solutions which could be more economically and socially acceptable in the immediate future? If this energy transition is desirable for the planet and humanity, could it be less brutal and benefit the planet's entire population, not to mention biodiversity?

These are the questions answered by the four authors I present in this book. They are physicists, ecologists, environmentalists, and economists who provide a necessary counterbalance to the assumed agreed-upon narrative. A very much appreciated counterweight that allowed me to make up my own mind.

Finally, I am neither an expert nor a scientist. But just like in 2006, I had sketched the anatomy of the virtuous so-called "Quebec model" to conclude that it was neither just nor equitable (but very costly), I wanted to delve into the climate issue.

Having introduced in chapter 1 the authors who inspired this book, chapters 2, 3 and 4 explore the state of climate science and the Intergovernmental Panel on Climate Change (IPCC). Is it warming or not? Is it dangerous or not? Does climate change really increase the severity and frequency of extreme weather events? What does the IPCC say? Finally, chapter 4 presents the fascinating world of climate models and scenarios. Is climate modeling even possible? Are climate models reliable? What do our authors think?

Once IPCC universe is explored, chapter 5 discusses the UN consensus culture and its climate apocalypse narrative. Is there really a consensus, and if so, what is it? What are the disagreements and debates about? A severe criticism is made of the agents propagating climate alarmism: media (reporters and publishers), politicians, activists, and some scientists. If there *is* a consensus, it seems to be on the poor quality of journalism and on a cataclysmic narrative reckless of consequences on mental health, on governments, on costs, and on the environment itself.

Chapters 6 and 7 then look at the politicization of the IPCC, its review processes, and summary reports that sometimes do not reflect the content of those they summarize. Putting aside the plan to reduce CO_2 emissions, we look at the United Nations' political agenda, that of the 'transformational changes' of our societies. Chapter 7 deals more specifically with the Net-Zero fad. Where do these 2°C and 1.5°C targets come from? Are they realistic? What do our authors think?

At the end of this chapter, we will be able to characterize the positions of three groups that speak on the subject: the CO_2-alarmists, the CO_2-lovers and the CO_2-realists. Here's a hint: there is rarely consensus.

Finally, chapter 8 provides an overview of the current situation, and makes the case that it's time to turn down the thermometer of climate doom and gloom.

So, it is with great pleasure that I invite you to read my "What I wanted to know about climate change".

CHAPTER 1

Climate disinformation: sure, but which one?

Are the lungs of the earth *really* burning?

In June 2020, the American Michael Shellenberger published a text entitled *On Behalf of Environmentalists, I Apologize For The Climate Scare.*[1]

From then on, this environmental and climate activist would no longer participate in climate misinformation. According to some surveys, a significant proportion of the population feared the extinction of the human race while others reported that young people between the ages of 8 and 16 had sleep disorders and nightmares about climate change.[2]

Enough! thought Shellenberger.

That same year, he published his essay *Apocalypse Never* (probably a reference to the movie title *Apocalypse Now*). In one of his chapters, Shellenberger attacks a strong and highly emotional image, namely that the forests of the Amazon are the 'lungs of the earth'. Indeed, who can remain insensitive when forest fires and deforestation occur in the Amazon for agricultural purposes? Add to that the power of a few activist stars and you find good reason to publicly denounce the policies of a newly elected Brazilian government perceived as being far-right.

This is what Leonardo DiCaprio, soccer star Christiano Ronaldo and Madonna did in August 2019 after flying over the forest fires that devoured part of the Amazon. On DiCaprio's Instagram and Ronaldo's twitter feed, this is what was shared with supporting photos,[3] writes Shellenberger:

> "The lungs of the Earth are in flames." - DiCaprio
>
> "The Amazon Rainforest produces more than 20 percent of the world's oxygen." - Ronaldo

Of course, the media picked up these tweets and added to them, seizing the opportunity to blame the new president of Brazil Jair Bolsonaro, even if the deforestation of the Amazon had increased at least *six years before* he came to power. Moreover, regarding wildfires, "while the number of fires in Brazil in 2019 was indeed 50% higher than the year before, it was just 2% higher than the average during the previous ten years,"[4] says Shellenberger.

A month later, Greta Thunberg and other student climate activists are suing Brazil. They believe that the country is not doing enough to slow climate change and according to them, the Amazon absorbs "a quarter of the carbon absorbed by all the forests on the planet". As for the President of France Emmanuel Macron, the house is burning!

> "Our house is burning. Literally. The Amazon, the lungs of our planet that produce 20% of our oxygen, is on fire. This is an international crisis."[5]

Michael Shellenberger has a particular interest in tropical forests. At the age of 16, he was raising money for the *Rainforest Action Network* which called to boycott Burger King who bought beef from Costa Rica's ancient rainforests. At 21, he attended the United Nations Environment Summit (Planet Earth Summit) in Rio de Janeiro, Brazil, where one of the hot topics was deforestation. He learned Spanish in Nicaragua and Portuguese. He lived in Brazilian communities that practiced slash and burn farming.

Let us agree that it would be very difficult to think of Shellenberger as a far-right-leaning person. Nonetheless, during the Amazon fires of August 2019, he contacted the lead author of an IPCC report on the Amazon, Dan Nepstad. "Is it true that the Amazon is a major source of Earth's oxygen supply?" he asks.

> "It's bullshit [...] There's no science behind that. The Amazon produces a lot of oxygen, but it uses the same amount of oxygen through respiration, so it's a wash."[6]

He continues his quest and approaches an ecologist from the University of Oxford for more explanations. The vegetation consumes about 60% of the oxygen it produces, says the expert. Biomass-breaking microbes consume much of the remaining 40%.[7] On August 26, 2019, Shellenberger writes about it in *Forbes*: *Why Everything They Say About The Amazon, Including That It's "The Lungs of the World" Is Wrong*.[8]

As for the photos posted by the stars on social media, it seems they were not photos of the August 2019 fires. Some were not even from the Amazon! "The photo shared by Ronaldo was taken in southern Brazil, far from the Amazon, and it was taken in 2012, not 2019,"[9] he tells us. Madonna's photo was over 30 years old!

Out goes the emotionally charged image of the Amazon as the lungs of the earth. Surely a very strong image but unfortunately the result of political disinformation and activism that do no justice to the environment or to the local populations.

Polar bears, victims of climate change?

The Danish Bjorn Lomborg is probably the one who has clashed the most and for the longest time with the environmental apocalypse supporters, especially since the 2001 release of his book *The Skeptical Environmentalist.*[10] Here's what the prestigious magazine *The Economist* had to say about it.

> "This is one of the most valuable books on public policy - not merely on environmental policy - to have been written for the intelligent general reader in the past ten years. ... The Skeptical Environmentalist is a triumph." [11]

Some 10 years later, in 2010, following the publication of another essay *Smart Solutions To Climate Change – Comparing Costs and Benefits*[12], Montreal's *La Presse* met with Lomborg and asked him "Have you changed your mind?" Lomborg:

> "Not at all. I always said that global warming was a problem and that it was man-made."[13]

With a title like *Bjorn Lomborg: un sceptique de moins* (one less skeptic), did *La Presse* welcome Lomborg's conversion? Did they not understand that Lomborg's scepticism was not about the recent global warming but about the CO_2 hysteria shown by governments, media, and activists? For example, in the preface to *Cool It* published three years earlier in 2007, we understand that Lomborg is far from denying global warming:

> "That humanity has caused a substantial rise in atmospheric carbon dioxide levels over the past centuries, thereby contributing to global warming, is beyond debate.

What is debatable, however, is whether hysteria and headlong spending on extravagant CO_2-cutting programs at an unprecedented price is the only possible response. Such a course is especially debatable in a world where billions of people live in poverty, where millions die of curable diseases, and where these lives could be saved, societies strengthened, and environments improved at a fraction of the cost."[14]

Lomborg is unsettling. Because for him, recognizing the role of CO_2 is one thing. Ways to limit warming and lessen its effects are another. Few people, especially activists and media, want to see a difference. Moreover, for Lomborg, there are alternatives to CO_2 reductions that could be more efficient.

He therefore advocates a change of tone and a robust cost/benefit approach that would support solutions that will benefit all, including developing countries. Radical solutions emanating from the Kyoto Protocol (1997) will not do, according to Lomborg. Neither does the Paris Agreement that will follow in 2015.

And so, in 2004, he creates the *Copenhagen Consensus Centre*, a *thinktank* that brought together economists and Nobel laureates, among other experts, to find smart, evidence-based solutions to climate change. "Data is my religion," says Lomborg in an interview with Canadian psychologist, author, and commentator Jordan Peterson.

"At present, anyone who does not support the most radical solutions to global warming is deemed an outcast and is called irresponsible and is seen as possibly an evil puppet of the oil lobby. It is my contention that this is not the best way to frame a debate on so crucial an issue."

We agree. But let's get back to our polar bears…

The image of a polar bear on a small piece of ice is surely one that disturbs and is very moving when it comes to associating climate change with the vulnerability of the species.

In *Cool It* and *False Alarm*, Lomborg recalls the heart-breaking image of an old bear making the cover of *Time*[15] magazine in 2006. "*Be Worried. Be Very Worried*" was the title of the special issue on global warming. That same year, Al Gore revived the iconic image in the film *An Inconvenient Truth*. In 2017, the prestigious magazine *National Geographic* even went with an image and a video showing an old hungry polar bear. It is such a success of environmental communication that it is the subject of an analysis in the journal *Environmental Communication*![16]

But what is really happening? With more summer ice melting in the Arctic, the *World Wildlife Fund* even warned that the species would be extinct by 2022! Lomborg in *Cool It:*

> "The World Wildlife Fund actually warns that polar bears might stop reproducing by 2012 and thus become functionally extinct in less than a decade. In their pithy statement, 'polar bears will be consigned to history, something that our grandchildren can only read about in books.'"[17]

Well, we're now in 2023, and according to zoologist Susan Crockford, who has studied the ecology and evolution of polar bears for over 40 years, there's no need to worry about these majestic beasts. In her book *Polar Bear Facts & Myths: A Science Summary For All Ages:*

> "There are many more polar bears in the world now than there were 50 years ago, when so many polar bears had been killed by hunters that people were right to worry about their survival. But new rules worked to protect the polar bear

> from too much hunting and at last count (2015), there were as many as 31,000 bears."[18]

What about the melting ice in the Arctic, then? Doesn't it make it harder for the polar bears to hunt seals? On her blog, Crockford writes (emphasis *hers*):

> "The evidence to date shows that declines in summer sea ice **have not** been detrimental to polar bear health and survival. On the contrary, in virtually all areas, much less sea ice has been beneficial: the opposite of what polar bear models predicted would happen."[19]

In a *Wall Street Journal* article, Lomborg is delighted with the growth of the polar bear population since the 1960s. A success story, according to him, which he also attributes to 1976 international agreement to regulate hunting.

> "Activists have used polar bears as an icon of climate apocalypse for decades, but the best data show that far from dying out, their numbers are growing. The official assessments from the leading scientists who study these animals—the Polar Bear Specialist Group within the International Union for Conservation of Nature—peg the global population today at 22,000 to 31,000. That's higher than the 5,000 to 19,000 polar bears scientists estimated were around in the 1960s.
>
> The main reason has nothing to do with climate. An international agreement enacted in 1976 limits polar-bear hunting, always the key threat to polar bears' numbers. Polar bears survived through the last interglacial period, 130,000 to 115,000 years ago, when it was significantly warmer than it is now."[20]

Isn't this fascinating? Why don't we hear this "good news"? Because that would contradict the official narrative that climate change threatens the existence of polar bears? What is so damaging about presenting different currents of thought? And is it not time to re-examine the relevance of spreading such emotionally charged images to young children in our schools?

The real threat to many species is not climate change, according to Lomborg. It's humans. "Still today, some 700 polar bears are killed by hunters each year. If we want to help polar bears, why not stop shooting them?" he concludes.

As for *National Geographic*'s old and starving bear, there is now a note from the publisher on their website:

> **Editor's Note:** National Geographic went too far in drawing a definitive connection between climate change and a particular starving polar bear in the opening caption of our December 2017 video about the animal.[21]

In August 2018, the CBC reported:

> "One year after a controversial video linked an emaciated polar bear to climate change, National Geographic has published an article on what went wrong."[22]

Climate disinformation

In *Unsettled*, Steven E. Koonin tells us that he always had this desire to understand the world and to measure what surrounded him. In kindergarten, he was fascinated by a thermometer that was in his class. Wanting to experiment with it, he observed changes in temperature by walking in the corridors, outside the school, and then at home. Having borrowed it without permission, he regretfully had to, with an apology, return the thermometer to school when his mother discovered it in the freezer at home.

A few days later, his parents got him one. This passage made me smile…

Koonin is a scientist. As a physician, he observes, measures and gets data. As a 'data-lover', he is probably the one of the four authors I introduce in this book who is the most outraged by how the IPCC reports are interpreted.

> "It's clear that media, politicians, and often the assessment reports themselves blatantly misrepresent what the science says about climate and catastrophes. Those failures indict the scientists who write and too casually review the reports, the reporters who uncritically repeat them, the editors who allow that to happen, the activists and their organizations who fan the fires of alarm, and the experts whose public silence endorses the deception." [23]

Here are some scary quotes on climate predictions[24] taken out from his book. Make a note of the dates.

- "[Inaction will cause] . . . by the turn of the century (2000), an ecological catastrophe which will witness devastation as complete, as irreversible as any nuclear holocaust." - Mostafa Tolba, Former Executive Director of the United Nations Environment Program, 1982.

- "[Within a few years] winter snowfall [in the UK] will become a very rare and exciting event. Children just aren't going to know what snow is." – David Viner, Senior Research Scientist, 2000.

- "European cities will be plunged beneath rising seas as Britain is plunged into a Siberian climate by 2020." – Mark Townsend And Paul Harris, Quoting a Pentagon Report in the Guardian, 2004.

For Koonin, the United Nations Secretary-General Antonio Guterres (the 'Highway to Climate Hell' one), the United States government, the IPCC, the media, environmental activists, and unfortunately some scientists all contribute to distorting the facts.

In his book, he describes what the IPCC is: an institution under the governance of the United Nations to which contribute a multitude of teams spread around the world. According to him, one of these groups, the Working Group 1 (WG1), the one that evaluates the scientific aspects of the climate, describes quite accurately the physics of the climate. As for the other groups, and particularly those who produce summaries for international and national policymakers, he is concerned that their work does not accurately report scientific observations.

Decisions that commit trillions of dollars to reduce the influence of human activity on the climate must be based on information specifying the degree of uncertainty, he tells us. Otherwise, politicians can all too easily exploit uncertainties.

> "And uncertainty can be a political weapon. Politicians on the right who deny even the basics that science *has* settled [...] are not above exploiting climate science uncertainties, offering them as "proof" that the climate isn't changing after all.
>
> Politicians on the left find it inconvenient to discuss scientific uncertainties or the magnitude of the challenge in reducing human influences. Instead, they declare the science settled and label anyone who questions that conclusion "a denier," lumping conscientious scientists advocating for less persuasion and more research with those openly hostile to science itself."[25]

Encouraging this misinformation often spread by reports of the IPCC themselves would be irresponsible, he says. Evidently, he is not happy with scientists who hide the limits of their models, researchers and activists who feed on biased-based funding, polarizing politicians playing on fear, and various journalists graduated in communication but unfortunately too click-dependent.

So, you may ask, how does he answer the question of podcaster Michael Shermer,* "What about the 97% of scientists who agree that global warming is man-made? That there is a scientific consensus?"

> "Pretty flawed. People speak out. But do I believe that the globe is warming: Yes. Do I believe that CO_2 is going up because of human activity? Absolutely. Do I believe that that exerts a warming effect on the globe? Absolutely.
>
> But do I believe that it's going to be catastrophic? NO! Why? Because I read the reports!"[26]

Having said that, for me, the most revealing part of this podcast was this one (emphasis is *mine*):

> "We're interested not in describing the climate, but **to describe how the climate responds to those human influences.** THAT'S the big question: a much more complex endeavour.
>
> We're looking at climate in a hundred years; yet we have terrible data to describe what happened in the past. The oceans: terrible data. But where the climate happens.
>
> **Untangling natural vs. human influences** is the central problem."

Listening to this podcast, I had just understood the magnitude of the problem. It was not enough to link CO_2 and global warming; *we also needed to assess all the other natural influences on the climate.* Only then could we assess the human influence or CO_2's contribution to climate change. No small feat! Among these influences, let us mention solar

---

* Michael Shermer is the founder of *Skeptic* magazine and the host-podcaster of *The Michael Shermer Show.*

intensity, the moon, the oceans, the clouds, the volcanic activity on land and in the seas, the effects of urbanization, ocean currents and so on. No small feat indeed.

A canonized consensus is not science

The climatologist Judith Curry presented earlier agrees. The climate system is extraordinarily complex, and isolating the effect of CO_2 is no small task. Unfortunately, the distraction caused by the so-called "*97% of scientists agree that climate change is due to human activity*" did not help.

In one of the first chapters of her book, Curry explains how this *so-called consensus* came about: from a 2013 paper by Cook et al. that analyzed the abstracts of 11,944 climate-related papers published between 1991 and 2011. To qualify, the words 'global climate change' or 'global warming' had to be mentioned in the abstract. Here is Cook study's conclusion taken from its own abstract.

> "We find that 66.4% of abstracts expressed no position on AGW (Anthropogenic Global Warming), 32.6% endorsed AGW, 0.7% rejected AGW and 0.3% were uncertain about the cause of global warming. Among abstracts expressing a position on AGW, 97.1% endorsed the consensus position that humans are causing global warming." [27]

And so, *by excluding the 66,4% of studies that did not take a position*, the authors concluded that 97.1% endorsed the idea that climate change was due to human activity! An advertising campaign followed, the approach was contested but for Cook, it was necessary that the public put pressure on their political representatives for something to happen! In an interview with *Scientific American,*[28] Cook explains the reason for this initiative.

For him, promoting a scientific consensus on anthropogenic warming could accelerate the implementation of action plans. It was necessary to emphasize the almost unanimous support of scientists and to counter the sceptics who according to him poisoned the minds of Americans.

It was indeed a very successful campaign that went viral, writes Judith Curry, especially when then US President Barack Obama tweeted in 2013 that:

> "Ninety-seven percent of scientists agree: #climatechange is real, man-made and dangerous."[29]

But President Obama's tweet is misleading, she says. It was an example of consensus entrepreneurship, but 97% of a subset of abstracts published between 1991 and 2011 cannot be interpreted as 97% of scientists. Nor did the paper refer to "dangerous" climate change". [30]

But even if this famous 97% has been much written about, was it *really* the most important thing to remember, Curry asks? (The emphasis is mine).

> "While this paper and its methodology were widely criticized, the important point is not whether the number is 97 percent or some other number. The key scientific issue is not *whether* anthropogenic greenhouse gases have caused any increase in global temperature – ***the issue is how much global warming has been caused by humans***, about which there is disagreement among scientists." [31]

Again, and as Koonin also points out, the real question is not *whether* GHGs influence climate* but *how much* human activity contributes to global warming. And for that, one must first understand direct and indirect natural influences on the climate; only then would we be able to distinguish between them and human activity influences. Clearly, we are not there yet. Believes Curry, when faced with such a complex and uncertain situation (climate assessment), the very concept of consensus is irrelevant. And yet,

> "The IPCC consensus has become canonized through a political process, bypassing the long and complex validation process for such a complex topic as to whether the conclusions are actually true. By representing their work as a consensus, the IPCC essentially delegitimized the objections of those commonly labeled as skeptics or deniers." [32]

For a specialist and analyst of uncertainties and risks related to climate change, and once a fervent defender of the IPCC, one can understand her discomfort with a narrative centered on a quasi-dogmatic consensus, as well as with some scientists' hypersensitivity to criticism. As for another controversial IPCC episode, that of Climategate in 2009, here is what she has to say about it:

> "A big part of my reaction to events unfolding after Climategate was the concern that I had somehow been duped into substituting the judgments of the IPCC for my own in my public statements on climate change. [...] When I read the Climategate emails, they confirmed concerns and suspicions that I had been developing about politics and

---

* Among the people quoted in this book, there is indeed a consensus on the contribution of CO_2 to atmospheric warming, but to what extent is still unknown.

personal agendas encroaching on the IPCC assessment process."[33]

The Climategate episode is certainly one of the most important controversies that has shaken the climate change scientific universe. It is about the scientist and very influential Michael Mann's famous *hockey stick* image. The same image you must have seen in Al Gore's film *The Inconvenient Truth*: a graph showing fairly stable temperatures from the last 1000 years until 1900, followed by a significant rise in temperatures since then.

However, it turns out that two Canadians[1] demonstrated that this was based on fundamental methodological errors. A short version of this story is available on YouTube[34] and is told by McKitrick himself. In summary, it took the U.S. Congress to make the study's code and data available. Two research committees were created to analyze Mann's team's work and, in the end, two reports endorsed the Canadians' findings. The data could not support Mann's conclusion, the methodology was incorrect, and as a bonus, the committees blamed Mann's team for working in silos and lacking transparency.

* * *

What is there to understand from all this? That in the coming years, we will hear more and more scientists speak out against misinformation and political intervention in the UN's and IPCC's climate science assessment process. The Amazon 'lungs of the earth', the starving polar bear, the 97% and *ClimateGate* episodes, the sweeping statements of political

[1] The Canadian economist Ross McKitrick, and the mathematician Steve McIntyre.

leaders and the media are only some of the unfortunate examples.

For these scientists, disinformation, the culture of consensus and eco-catastrophism help neither the evolution of science nor the population to see things more clearly. They must be reported and denounced.

Finally, I leave here two wonderful quotes borrowed from Curry's book that could reflect on the state of climate science today.

> "Science is always sold as facts, and it's not, it's process. And that process is mainly arguing." – Tamsin Edwards, Climate scientist.
>
> *"When you mix politics and science, you get politics."* – John M. Barry, author of *The Great Influenza.*

CHAPTER 2

The science of climate change

Climate change

If there is anything beyond dispute, it is that the planet's climate has always been changing. Glacial and interglacial periods testify to this. Here is a striking illustration of sea level changes over the last 400,000 years, taken from Koonin's *Unsettled,* p. 150.

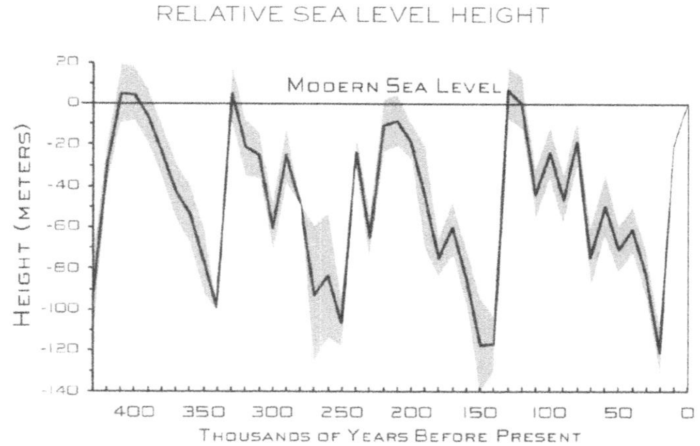

FIGURE 8.1 **Global sea level over the past 400,000 years as estimated from geological proxies.** The typical uncertainty in these estimates is 10 meters.[2]

During these long cycles, large temperature oscillations* cause ice sheets to either grow or melt. Therefore, depending on the period, the sea level goes down or up. Fascinating,

* Oscillations explained by variations in the earth's orbit and axis (rf. Milankovitch cycles)

isn't it? And what is even more fascinating is that it happens every 100,000 years... and by the way, even in the absence of any CO_2-generating human activity. Also interesting is what was happening 125,000 years ago.

> "During the last interglacial (low-ice) period 125,000 years ago, known as the "Eemian", sea levels were some 6 meters (20 feet) higher than they are today!"[1]

As to where we are now, here is Koonin's answer:

> "The Last Glacial Maximum occurred about 22,000 years ago, when the continental glaciers started melting once again. [...] Since then, sea level has risen by about 120 meters (400 feet) very rapidly until about 7,000 years ago, when the rate of rise slowed dramatically; that's shown in Figure 8.2 below."[2]

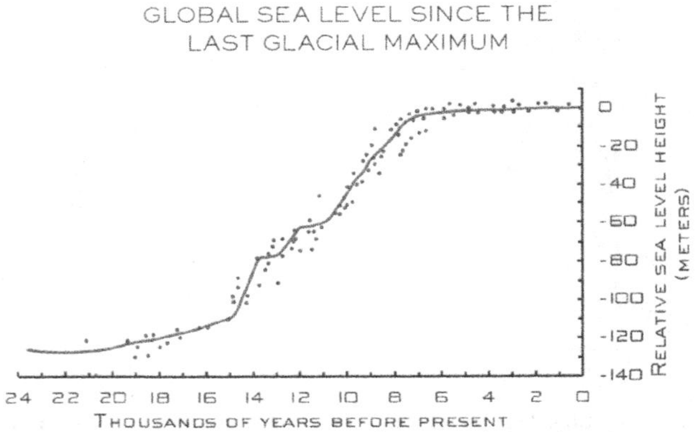

FIGURE 8.2 **Geological estimates of sea level change since the Last Glacial Maximum 22,000 years ago.** The solid curve is the average of estimates from various sites around the globe, shown by the individual representative points.[3]

And this interglacial era called the Holocene in which we are at present, would have started about 12,000 years ago. Thus, in this period spanning over tens of thousands of years, and as evidenced by the graph above, we would be at a time when the rising waters have "slowed down dramatically."

Climate vs Weather

Before we go any further, **climate is not weather**. In fact, according to the United Nations' World Meteorological Organization (WMO),

> *Climate is defined as a thirty-year average of weather conditions of a particular location.*

The climate of a region is therefore the average of temperatures measured over several decades. Koonin explains:

> "It takes at least a decade of observations to define a climate, and so two or more decades to identify a change in it."[3]

When extreme weather events occur, news reports therefore *incorrectly* associate these phenomena with climate change. "With climate change, we will have more and more floods, hurricanes, forest fires, drought," they say.

And then there are climate media activists that add to the scare. Here is an example of a *La Presse* columnist after the floods in Baie-St-Paul, Quebec, on May 4, 2023.

> "We won't escape it: there will be more and more weather reports on heat waves and torrential rains. The storm of the century will become the storm of the year. Roads will collapse. Ecosystems will be disrupted. People will literally die of heat."[4]

"I am simply repeating what the climatology experts have been telling us for years," she adds.

Not exactly, would reply Koonin. Meteorologists and the media err and misinform when they associate a particular weather event with climate change. Indeed, as we will see later, according to Koonin and several other scientists, the data reported by the IPCC group that is assigned to the scientific aspects of climate (the WG1) in no way confirms an amplification of all extreme climate events.

Is there global warming or not?

But is global warming really happening? Yes. Absolutely. It happens. All the time. And more recently, there has been a warming of 1.1°C since the beginning of the pre-industrial era (since 1850). Look at this chart which shows not temperatures but anomalies, that is, temperature deviations from the 1951-1980 average (this is how an anomaly is calculated and how we can see upward or downward trends).

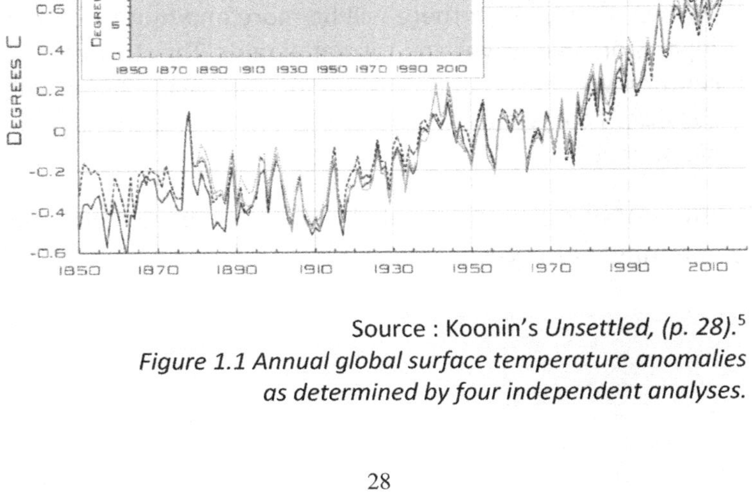

Source : Koonin's *Unsettled, (p. 28).*[5]

Figure 1.1 Annual global surface temperature anomalies as determined by four independent analyses.

Looking at this graph, we can clearly see that from 1900 to 2020, anomalies* went from -0.3°C to +0.8°C, showing an increase of +1.1°C over a 120-year period.

Interestingly, Koonin also points out that within this period there was also a cooling period, between 1940 and 1980. A cooling of -0.05°C per decade.

In fact, as recounted in a 2017 article *In 1975, Newsweek Predicted a New Ice Age*,[6] several mass media outlets in the 1970s were predicting global cooling! A new ice age! Mainstream publications such as *Time, The Los Angeles Times, Fortune, The Chicago Tribune, The New York Times*, and *National Geographic* heralded a "*Cooling World*."

Was it really the case? Yes. During the period 1940-1980, the planet cooled down a bit. But the idea that the planet would enter a new ice age was not the opinion of all climate researchers of the time. As the story goes, even if cooling was recognized in these few decades, some researchers persisted in the idea that in the medium term, the planet was warming up. How to explain the opinion of those who thought we were about to experience a colder period?

> "The study of the world's climate was still primitive in the 1970s. Few meteorological scientists then knew how to interpret trending temperature information, and the cause of climate changes was mysterious. The information that climate researchers had collected was incomplete and easy to misread."[7]

---

* Anomaly = Deviation from a baseline (average) value. Here, the baseline is the 30-year average of 1951-1980 temperatures. The four sets of values come from four different and independent methods of analysis. The differences between them are too small to be visible.

Let us remember here that climate "science" was primitive in the 1970s… And let us also remember that it can be tempting to choose certain time intervals rather than others to give hoped-for results and/or make projections à la Nostradamus.

Another indicator of global warming these years? According to Koonin, ocean warming, and this, even though science is still young.

> "Satellites can measure the temperature only at and above the ocean surface, and they have been doing even that for less than half a century."[8]

According to him, "there is a clear upward trend in the oceans' heat energy," a good indication that the planet is indeed warming.

> "[...] we're confident that the oceans have been warming over decades, if not centuries. [....] The world's oceans are both the most important and the most problematic piece of the earth's climate system. They hold more than 90 percent of the climate's heat and are its long-term memory."[9]

Settled? That the planet warms up, certainly. But can we assert that the cause is mainly human activity (mainly CO_2)? For this, Koonin says, it would be necessary to consider other studies, one of which concludes that the upper 2,000 metres of ocean surface had warmed between 1750 and 1950,[10] a time when human influences were very limited.

To sum up, it looks like science is indeed still young and immature to assert beyond any doubt the role of human activity and of CO_2.

More distant periods of warming

With regard to global warming, let us also mention that periods of significant warming may have occurred in an even more distant past.

For this, we must resort to paleoclimatology, a science that makes it possible to reconstruct the anomalies (variations) of the past climate using proxies.[**]

Here is an excerpt from a presentation of Patrick Moore, co-founder of Greenpeace and now director of the *CO_2 Coalition.*[11***] The small circle at the end of the graph illustrates where we are now, and where the climate is warming again.

[**] Proxy data are preserved physical characteristics of the environment that can stand in for direct measurements. Examples : tree rings, ocean and lake sediments, ice cores, etc.

[***] The *CO_2 Coalition* is a group of 120 scientists and researchers who recognize that the planet has been warming since decades but who think that CO_2 has very little to do with it.

Examples of warming periods in the last 5,000 years

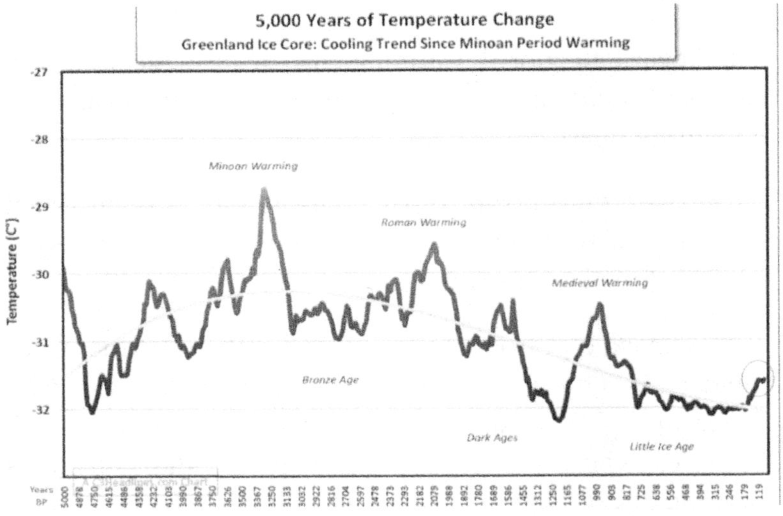

We can see here the significantly warmer periods that have occurred in the last 5000 years. If we look at the Minoan period for example (2700-1200 BC), some archaeologists have recently been suggesting that it was sudden changes in the climate that wiped out this civilization around the 1200s BC. Here is a quote by archaeologist Eric H. Cline:

> "A recent core sample analysis taken from the Sea of Galilee by Dafna Langgut and Israel Finkelstein of the University of Tel Aviv and Thomas Litt of the University of Bonn, irrefutably proved that the period between 1250 and 1100 BC was the driest ever observed during the Bronze and Iron Age."[12]

Cline believes that extreme weather events were important factors in the collapse of civilization at the time. He explains this in his book *1177 BC: The Year Civilization Collapsed.* In an interview with *World History Encyclopedia:*

> "There is both direct and circumstantial evidence of climate change, drought and famine, earthquakes, invasions and internal rebellions, all of which occurred around this time."[13]

This goes to show, says Cline, that profound changes in climate can upset entire populations and cause significant migratory movements, going so far as to annihilate an entire civilization, particularly if these changes are made over a relatively short period. He adds:

> "It would be difficult to survive if all, or most, of the above calamities came at the same time or nearly so, as they seem to have done especially between about 1225 BCE and 1175 BCE. And that, I think, is why the Late Bronze Age civilizations came crashing down—they were not able to weather the 'perfect storm' of nearly simultaneous catastrophes [...]."

The medieval period and the *hockey stick*

One of the other climate anomalies is that observed in the years 950-1250 of our era, during the medieval period. Here again, several scientific reports try to demonstrate the phenomenon, others try to invalidate it. Finally, some even claim that the IPCC goes so far as to reject reports that designate this period as one of *global* warming. Why, would you ask? One theory is that if such warming could occur in the past at a time when CO_2 could not explain it, it would destabilize the thesis that current global warming is entirely due to human activity.

But if you want to get to the bottom of it, I wish you good luck because there is a vigorous debate about this warmer medieval period (called MWP for Medieval Warm Period).

To make a long story short, some scientists characterize this particularly warm period as *global* warming, while IPCC reports downplay its importance by arguing that these were *regional warming* changes and that they were limited to the European North Atlantic. No, other researchers reply. The warmer period was also happening in the Antarctic Peninsula, the Andes, North America, eastern Africa, China and New Zealand.[14] This was followed by a cold period (1450-1850) known as the *Little Ice Age*, when temperatures were the coldest in 10,000 years.*

Again, why so much debate on whether or not global warming has occurred before the pre-industrial era? Because if it were proven that there already were such periods in the past (and therefore before the use of fossil fuels), it would be more difficult to conclude that the warming observed today is essentially linked to CO_2 emissions. The IPCC's hockey stick thesis could be called into question. According to some, it is misleading and makes politicians and the public believe that our time is exceptional and essentially the result of human activity. Here, as an example, is an excerpt from an article *Who Erased The Medieval Warm Period*. The author is exasperated by the most recent IPCC assessment report (the 6th of 2021) which, once again "makes the warm medieval period disappear." It recalls the introduction of the famous hockey stick in the 3rd IPCC assessment report in 2001.

> "The 3rd Climate Assessment Report of 2001 already contained a similar field hockey stick pattern, intended to

---

* It seems to me that the IPCC and those who are worried about global warming stipulate implicitly that the ideal temperature would be those of the pre-industrial era. Tell me, who would want to return to temperatures that were the coldest in the last 10 000 years? Odd, isn't it?

fool politicians into thinking that today's warming was unprecedented and therefore entirely man-made."[15]

The " hockey stick"

From a 1988 paper[16] by three authors, among which we find Michael E. Mann

Demonstrates with reconstruction techniques that temperatures were relatively stable since the early 1400s after which there is a significant increase in temperature starting in the mid-twentieth century. Interestingly enough, from then on, Mann argued that greenhouse gases (GHGs) were the determining factor in climate warming.

The "hockey stick" was first introduced in the 3rd IPCC Assessment report in 2001. To this day, still challenged by many researchers and scientists.

Is 'science settled'? It seems that many scientists do not believe so and continue their research, which in my opinion is the characteristic of science. For them, erasing global warming periods such as that of the medieval period is not based on reliable scientific methods of reconstructing past climate episodes.*

As for Koonin, here is an image from his book taken from the 5th IPCC Assessment Report of 2013 (AR5). Look carefully at a reconstruction of the temperature variation since the year 500 (compared to the average of the period 1881-1980). And note that in 2013, the IPCC *does* recognize

* A book entitled *A Disgrace to the Profession* by Mark Steyn, 2015, is a collection of testimonies from many scientists who deplore Michael Mann's methods in rebuilding the hockey stick (some of whom adhere to the consensus that there is, in fact, global warming).

the higher temperatures around the year 1000 (warmer medieval period) that were followed by the Little Ice Age.

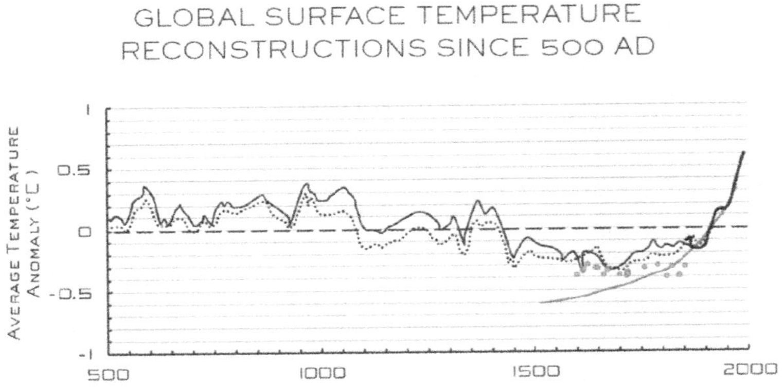

FIGURE 1.7 **Global average surface temperature anomalies during the last fifteen hundred years as reconstructed by different proxy methods, together with the modern instrumental record (black line).** Anomalies are relative to an 1881–1980 baseline and have been smoothed to reduce variations on timescales shorter than about fifty years.[15]

Source: Koonin's *Unsettled, p. 41*[17]

> "As you can see, centuries of warmer temperatures gave way gradually around the year 1000, leading up to the Little Ice Age, an unusually cold period that ran from about 1450 to 1850. [...]
>
> Because of the limited evidence, the most recent assessment report (AR5) has only *low confidence* that the global warming of the past thirty years has exceeded the range of reconstructed temperatures."[18]

But then... pouf! In 2021, the medieval period is no longer noticeable in the IPCC Report. Here is the graph that appears in the 6th IPCC's *Summary For Policymakers.*[19]

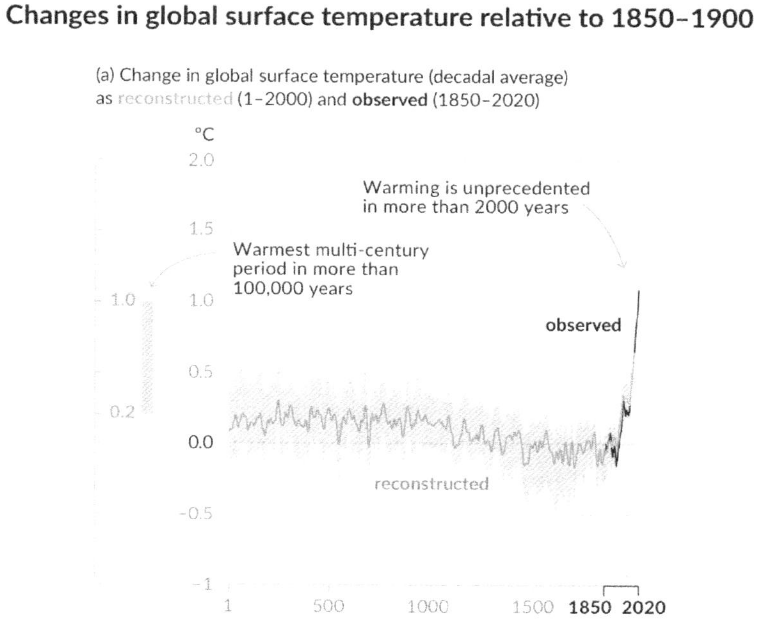

For the IPCC, what we need to understand is very clear: "*Human influence has warmed the climate at a rate that is unprecedented in at least the last 2000 years.*"[20] This is serious…

One thing for sure, research will continue, either within the IPCC or otherwise. Perhaps one day we will get closer to the truth with the contributions of other scientists such as archaeologists, paleoclimatologists, historians, etc. And perhaps reconstructions of the past will rely on techniques that will be more difficult to refute. For now, I postulate that science itself is under construction.

What can we expect in the future?

Finally, what does the IPCC predict for the coming decades? Well, the IPCC does not make predictions per se. What they do is establish scenarios of potential futures. A later chapter presents these scenarios as well as what some critics have to say about them. In my opinion, they range from 'candy pink' over-optimistic scenarios to scenarios so pessimistic that they are not even credible.

In conversation with Jordan Peterson, Koonin responds the following to the question, "Why should we not *act urgently* on global warming? Are we not hearing that if we delay action, suffering will only be worse in the future?"

> "Nonsense! That's NOT what the science says. Let me start with something relatively simple. Let's start with the last 120 years. In that time, we've seen the greatest improvement in human betterment ever seen.
>
> The population has gone up by a factor of 5. The GDP per capita has gone up by a factor of 7. Longevity increased from 32 years to the current 72 years. [...]
>
> To think that an additional warming is going to reverse or even significantly derail the progress we've had is just nonsense. So, the notion of a climate catastrophe is just nonsense."[21]

Clearly, for Koonin, the notion of climate catastrophism is absurd. In the last 100 years, mankind's quality of life has improved like never before, despite the addition of 1.1° C. To claim that another degree will turn everything upside down is pure madness. And why is he so sure about this? "Because I read the reports!"* And when you read about extreme

---

* By "report" Koonin refers to the Working Group 1 reports. As for the Summaries for the Policymakers, that's another story.

weather events, it's light years away from what you hear from activists, media and governments.

Let's summarize :

- Climate change exists. Since forever.
- The latest news is that the planet has warmed by about 1.1°C since 1900.
- Repeatedly, major fluctuations cause upheavals, whether they are unfortunate (migration, drought, etc.) or beneficial (greening of forests, more suitable land for agriculture, food security, etc.).

Science and the United Nations Framework Convention on climate Change (UNFCCC)

In 1992, Brazil hosted the 3rd Earth Summit. One of the Rio Janeiro's agenda was to establish a United Nations Framework Convention on Climate Change. The ultimate objective is crystal clear (Bold characters are mine).

> "The ultimate objective of the Convention (...) is to **stabilize greenhouse gas concentrations** 'at a level that would prevent **dangerous anthropogenic** (human induced) **interference** with the climate system'."[22]

This paragraph is dense and important. Not only does it define the objective (the protection of the climate system for present and future generations), but it also imposes the means (stabilize GHGs), determines the cause of these GHGs (anthropogenic disturbance) and asserts that these disturbances are dangerous.

As author Bernie Lewin recounts in his book *Searching for the Catastrophe Signal* – (yes, I got carried away with my readings...) it was no wonder that the Secretary-General of the Summit, a man named Maurice Strong, opened the event with a sense of urgency:

> "The IPCC has warned that if CO_2 emissions are not cut by 60 percent immediately, the changes in the next 60 years may be so rapid that nature will be unable to adapt and man incapable of controlling them."[23]

Towards the end of the Summit, he writes, several delegations from around the world had signed the text. The Convention entered into force in 1994 and the first Conference of the Parties (COP1) was announced in late March 1995 in Berlin. A great success.

But in what state was climate science when the Convention was signed?

At that time, the first IPCC climate assessment report of 1990 was available. Preceding that, in 1988, a "Conference in Toronto on the Evolution of the Atmosphere" had been held. Forty-six countries participated, as did their government representatives. One of the recommendations linked global warming to CO_2 emissions:

> "Set energy policies to reduce the emissions of CO_2 and other trace gases. Stabilizing the atmospheric concentration of CO_2 is an imperative goal. It is currently estimated to require reductions of more than 50% from present emission levels."[24]

According to Lewin that tells the story of the IPCC, "this rather arbitrary target stuck."

Curry agrees. According to her, the premise of the UN Convention was that global warming was *man-made and dangerous.* To "control the climate," we had to eliminate emissions. Simply put, disaster awaits if we do not reduce fossil fuel emissions. And all this at a time, in 1992, when there was *no scientific basis* on which to support these premises.

In terms of "detection," i.e., proof that warming is scientifically measurable, the first IPCC report of 1990 is still a long way off. Much pressure had been put on the IPCC team to find evidence that would confirm this detection, but on the eve of the Rio de Janeiro Summit, the authors remained cautious. An update of the first report had even been prepared in 1992 with a Supplement specifically about detection.

But in the end, the Supplement explains that "progress is difficult," and there were unresolved problems with models and so projections were in doubt. Basically, it was clear that the climate system responded to many different influences; but could GHG influences be distinguished from other influences?

As for the dangerousness of anthropogenic influence, the IPCC was faced with a dilemma. Lewin:

> "The IPCC's dilemma was this : it could not yet say *scientifically* what was 'dangerous anthropogenic interference' because as yet it had no *scientific* confirmation of any interference at all. In the meantime, the nations could choose to act without waiting for the science to catch-up, *which was entirely up to them.* But this would be a *political* decision. For Bolin (the author's report) without a shift in the science, to *answer this question of what is 'dangerous' would be to step into the political."*[25]

A step that he did not want to take.

So, this is where the IPCC research was at the time of the Earth Summit in Rio de Janeiro. Detection was not there and for the concept of dangerousness, it was considered that the question fell within the political domain.

> *In other words, at the time the United Nations Framework Convention on Climate Change was written and signed, there was no science to demonstrate global warming, its causes (GHGs) or its dangerousness ...*

In Lewin's words (and I read this many times to be sure I understood correctly):

> "This is how we have an assessment of the science behind manmade climate change published especially to support discussions of a treaty to mitigate manmade climate change that declares no improvement in the validation of the models predicting the change to be mitigated."[26]

That said, the Rio Earth Summit and the UN Framework Convention were a political success. "The political kart was way ahead of the scientific kart from the very beginning," Curry says in an interview with Peterson. "This convention was signed before anything was confirmed scientifically." As Spock would say, "Fascinating…" (Star Trek).

As for the premise of dangerousness, can we really attribute extreme weather events to human activity and CO_2? Let us look at what the IPCC has to say about it in its most recent report.

CHAPTER 3

Extreme meteorological events

Climate disasters

In August 2021, this is what UN Secretary-General Antonio Guterres had to say on the most recent IPCC's *Assessment Report (AR6)*.

> "The alarm bells are deafening, and the evidence is irrefutable: greenhouse-gas emissions from fossil-fuel burning and deforestation are choking our planet and putting billions of people at immediate risk.
>
> Greenhouse-gas concentrations are at record levels. Extreme weather and climate disasters are increasing in frequency and intensity."[1]

In Septembre 2022, this is him again as reported in a United Nations press release entitled *Secretary-General Warns Climate Crisis Heading Into "Unchartered Territories of Destruction"*.

> "Floods, droughts, heatwaves, extreme storms and wildfires are going from bad to worse, breaking records with ever alarming frequency. [...]
>
> There is nothing natural about the new scale of these disasters. They are the price of humanity's fossil fuel addiction. The number of weather, climate and water-related disasters has increased by a factor of five over the past 50 years."[2]

We must admit that the press releases' headlines are catchy! Enough to be used as is by the media, without any nuance or counterweight. Many frightening words in these statements: code red, state of climate crisis, the planet is suffocating, billions of people at risk, unchartered territories of destruction, alarming report… Nothing here to calm down eco-anxiousness and nuance what young people are being taught in our schools, it goes without saying.

Notice how the Secretary insists on the increasing number of climate disasters. They would be *more numerous and the costs would increase*. There is "nothing natural" in what we observe in terms of severe climatic events. And yet… There is nothing so dramatic in the IPCC Reports.

Koonin in *Unsettled*:

> "Indeed, it has become *de rigueur* for the media, politicians, and even some scientists to implicate human influences as the cause of heat waves, droughts, floods, storms, and whatever else the public fears. […] But the science tells a different story.
>
> Observations extending back over a century indicate that most types of extreme weather events don't show *any* significant change – and some such events have actually become less common or severe – even as human influences on the climate grow. In general, there are high levels of uncertainty involved in detecting trends in extreme weather."[3]

Shellenberger in *Apocalypse Never:*

> "Anyone who believes climate change could kill billions of people and cause civilizations to collapse might be surprised to discover that none of the IPCC reports contain a single apocalyptic scenario. Nowhere does the IPCC describe developed nations like the United States becoming a

"climate hell" resembling the Congo."[4]

Lomborg in 2021 in a *Financial Post* article *IPCC Global Warming Report is more chill than you have read:*

> "The Intergovernmental Panel on Climate Change (IPPC) just released its latest climate report, and reactions from politicians and media pundits could not have been more predictable. Fitting the apocalyptic narrative many have spun lately, the environment editor of *The Guardian* summarized this scientific report as finding mankind "guilty as hell" of "climate crimes of humanity." (Needless to say, the report says no such thing.) [...]
>
> Of course, the UN has a long history of claiming catastrophe is right around the corner: The first UN environment director claimed half a century ago that we had just ten years left, and the then-head of the IPCC insisted in 2007 that we had just five years left."[5]

Curry, in *Climate Uncertainty and Risk:*

> "Among the greatest concerns about climate change are its impacts on extreme events such as floods, droughts, wildfires, and hurricanes. However, there is little evidence that the recent warming has worsened such events.
>
> The first half of the twentieth century had more extreme weather than the second half, when human-caused global warming is claimed to have been mainly responsible for observed climate change.
>
> Regarding the perception that severe weather events seem more frequent and more severe over the past decade, there are several factors in play. The first is increasing vulnerability and exposure associated with increasing population and concentration of wealth in coastal and other disaster-prone regions. The second factor is natural climate variability."[6]

On the subject of extreme weather events, I must introduce you to a long-time expert very much appreciated and quoted by those who really want a fair idea of what the IPCC WG1 climate assessment reports say (which can be very different from what we may read in the Summary Reports for governments and policymakers).

Roger Pielke Jr.[7] has been on the faculty of the University of Colorado Boulder since 2001. He has been working on issues of climate science and policy for over 30 years. He is not a stakeholder in the IPCC and can therefore conduct independent analyses. Like Koonin, he scrutinizes the work of the IPCC. To the delight of those interested, he is an extremely popular speaker and publishes on his Substack* to which I subscribe.

When the UN Secretary said that "the number of weather and climate disasters has increased by 500% in the last 50 years and billions of people were at risk," Pielke firmly reacted. He is by no means a climate denier, but when the United Nations or the IPCC adopts an alarming tone, on the verge of disaster, he is appalled and worried about the perverse effects on the institutions' credibility.

> "Not only is this wrong, but it is also irresponsible. Nowhere does the IPCC report say that billions of people are at immediate risk."[8]
>
> "As I and others have documented, Guterres' claim of an 500% increase in disasters is pure misinformation. You will never find a more obviously and egregious wrong claim in public discussions from a more important institution."[9]

---

* Substack: Publication platform for content providers

Because you see, the data needed to assess the number and damages of extreme weather events are rigorously compiled in what is called the EM-DAT, the *International Disaster Database*[10] of the Catholic University of Louvain in Belgium. Here, from the database, is what Pielke has made available on his *Substack.*[11]

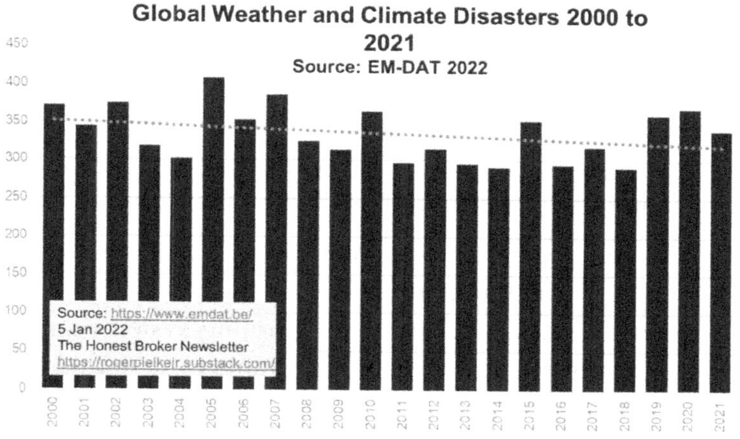

"Pielke's Weekly Memo #12," *The Honest Broker's Substack,* September 16, 2022

"Good news!", he writes. "So far this century, disasters are down. That trend continues well into 2022."[12]

Lomborg agrees. The number of deaths due to extreme weather events has not increased. In fact, he says, if we went back to 1920, the number of deaths even fell sharply.

> "In the 1920s, these disasters killed almost half a million people each year, mostly in large floods and droughts in developing countries. [...] Over the past hundred years, the number of deaths from these climate-related catastrophes has plummeted by 96 percent. Remember that over the same time period, the global population has increased fourfold."[13]

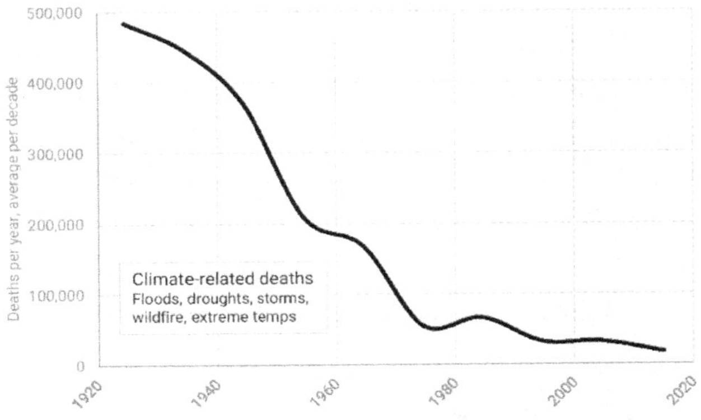

Global climate-related deaths from disasters, 1920-2019, averaged over decades – Taken from *False Alarm,* EM-DAT[14]

"This is a remarkable success story", says Pielke, considering the overall population increase during this period. A history of resilience due to the success of warning systems and the adaptation of populations more vulnerable to such events. A story that will not make the headlines but that could reassure some eco-anxious people sensitive to the evening news.

But what about costs? Indeed, if the number of events and deaths is going down, what about the astronomical costs of climate change damage? Is that not a reason to associate climate change with the emission of greenhouse gases and to expedite their reduction? Pielke has been studying extreme events' costs for several decades and major insurance companies are following this very closely as well. We are talking here about the costs and losses resulting from the destruction of assets (we understand the interest of companies in reinsurance).

We hear that damage costs are increasing in absolute numbers. But of course, they are increasing in absolute numbers! In his book *False Alarm*, Lomborg illustrates the rising cost phenomenon with a hurricane that hit Miami in 1940, reaching 24,000 homes. Today, the same hurricane would hit more than a million homes; still, the media will report the news in a way that makes us believe that events are more frequent and extreme.

However, if we consider that the global economy has doubled in the last three decades, losses due to meteorological disasters would have fallen from 0.25% of global GDP to less than 0.20% of global GDP. See the downward trend below.[15]

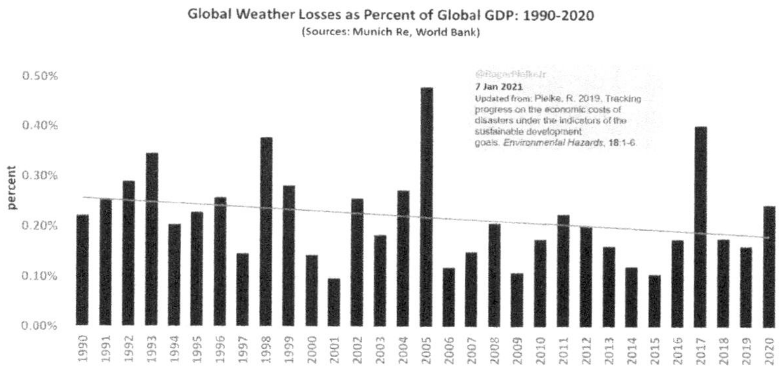

What can we learn from this? That neither the number nor the costs of weather disasters (relative to GDP) are growing.

The media *clickbait* business can certainly explain the catastrophist headlines. We cannot change that. Fear pays off. But when the UN Secretary-General, satellite institutes, researchers (funded and subsidized to support the official narrative) and even scientists sound the alarm on the grounds that the social and economic costs of climate disruption are more important, why not mention the real impacts of extreme events? Why does media deprive readers and listeners of a minimum of nuance and of balance in news coverage?

These sources are not credible, they may tell us. Simply misinformation from deniers? Not so. Still, reminding you that none of the five people I quote in this book is a climate denier, let's look at what the IPCC Task Force "WG1" has to say.

What does the IPCC say?

First, a definition of terms. Let us recall how "climate change" is defined in the United Nations Framework Convention.

> "'Climate change' means a change of climate which is attributed directly or indirectly to human activity that alters the composition of the global atmosphere and which is in addition to natural climate variability observed over comparable time periods."[16]

With regard to "detection" and "attribution," we can refer to the IPCC glossary,[17] but here is a more comprehensible version. Notice that we do not necessarily associate the 'detection' of a change of climate with human activity. This requires an 'attribution' process.

> "Detection consists in showing that the climate has changed according to certain defined statistical criteria, without giving the reason for this change. Attribution, on the other hand, involves establishing, with some degree of confidence, the most probable causes of the detected change. In other words, detection and attribution processes attempt to distinguish between natural and human influences on climate."[18]

Finally, the IPCC establishes degrees of certainty (from exceptionally unlikely to virtually certain) and confidence (very low to very high) to qualify its detection and attribution statements. Indeed, so many nuances that are rarely reported in the media.

Now here is an excerpt taken from Pielke's own synthesis of the IPCC report (AR6). The second column indicates whether, for the phenomenon in question, a change has been statistically observed in the past. The third column indicates whether this change can be attributed to human factors (CO_2).

Event	**Detection of trend**	**Attribution of trend**
Heat waves	Yes	Yes
Heavy precipitations	Yes	Yes
Flooding	**No**	**No**
Meteorological and hydrological drought	**No**	**No**
Agricultural drought	Yes	Yes
Tropical cyclones	**No**	**No**
Extratropical cyclones, hurricanes, tornadoes	**No**	**No**
Winter storms	**No**	**No**
Hail	**No**	**No**
Extreme winds	**No**	**No**
Fire weather	Yes	Yes

Taken from *How to Understand the New IPCC Report Part 2, Extreme Events*
Roger Pielke's "The Honest Broker" Substack[19]

According to the IPCC's conclusions, it is somewhat risky (and incorrect) to claim that the frequency and severity of certain weather events have increased, whether it be floods, meteorological or hydrological droughts, hurricanes, tropical cyclones, thunderstorms, hail, or high winds. Surprising, isn't it?

Indeed, it seems that a thorough read of IPCC reports makes it difficult to not come up with some observations of our own. For me, this is what Pielke's analysis show.

- That we really need to be mindful of the varying degrees of confidence and probabilities attributed to IPCC statements. Rarely is this piece of information mentioned in the media coverage.
- Several statements are understood by the media as being statements that concern the entire globe. However, they are often not. Such statements may apply only to certain parts of the world or to certain regions. Moreover, in some regions, there is not even enough observational data to conclude anything.
- It seems that climate change is always reported by the IPCC and the media as having disastrous consequences. But as Pielke reports, "an extreme weather event is not a disaster. A disaster exists only when such an event occurs in unprepared, vulnerable and exposed communities."[20] The level of wealth of a region is indeed a decisive factor in its ability to adapt to extreme weather events.

As a result, it became increasingly difficult for me to understand and even endorse the concept of "global climate". I also understood very clearly that if it's one thing to acknowledge change based on observations, it is another to attribute a specific change to human activity (CO_2).

The IPCC on heat waves

Let us take heat waves for example. This is what an IPCC statement looks like. Bold characters are mine.

> "It is *virtually certain* that **hot extremes (including heatwaves) have become more frequent and more intense** across most land regions since the **1950**, while **cold extremes (including cold waves) have become less frequent and less severe, with *high confidence* that human-induced climate change is the main driver of these changes**.
>
> Some recent hot extremes observed over the past decade would have been *extremely unlikely* to occur without human influence on the climate system."[21]

So much for the overall picture. Now, let us see how one cannot generalize for the entire planet and notice the importance of the time interval used.

First, the time interval used by the IPCC starts in 1950. Intriguing... because if there were any heat waves in the United States, it was around the 1930s. In fact, so severe was this period of heat that it bears a name: the *Dust Bowl.* Why start in 1950, then?

What follows are graphs that show the frequency (top) and intensity (bottom) of heat waves in the U.S. since 1900. They are taken from the Fourth US National Climate Assessment.

Notice in the top chart that the number of heat waves in the United States has been on the rise since the 1970s but barely reaches what was observed at the beginning of the last century in 1900. Also note the high intensity of heat observed around the 1930s.

Frequency (top) and intensity (bottom)
of heat waves in the US since 1900.
Taken from "What the media won't tell you about heat waves," The Honest Broker's Substack[22]

Source: Fourth U.S. National Climate Assessment. Source[2]; USNCA 2017

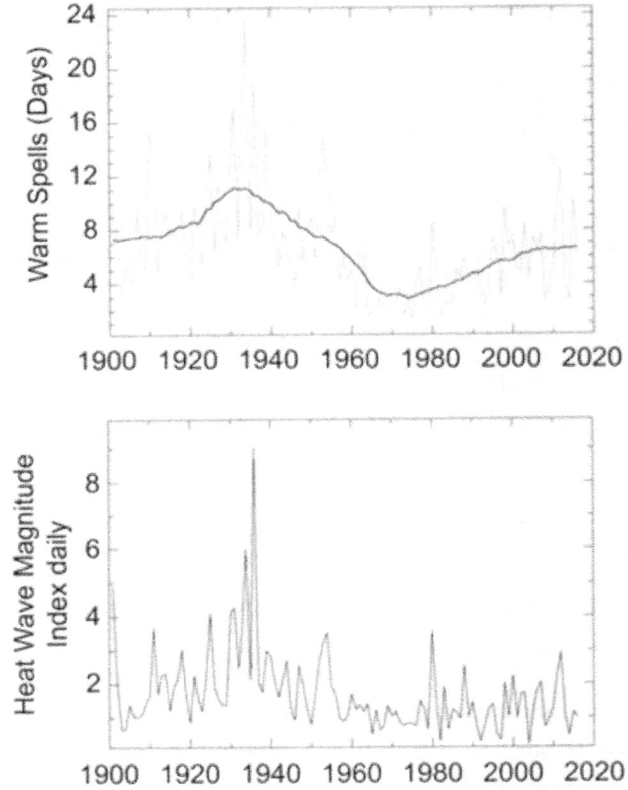

"The extreme temperatures of 1930s present a challenge for the detection and attribution of trends in heat waves in the United States,"[23] writes Pielke. "It is true that if analysis of

[2] This climate report is a product of the U.S. Global Change Research Program, mandated by the U.S. Congress.

data begins in the 1960s, then an increase in heat waves can be shown. However, if the data analyses begin before the 1930s then there is no upwards trend, and a case can even be made for a decline. It is a fertile field for cherry pickers," he says.

The other question is, how can you attribute an increase in heat waves to human activity when you remove those of the 1930s? How could we 'attribute' to human activity (and therefore, CO_2) the *Dust Bowl* years when CO_2 emissions were only one-tenth of those of today?

Can we suspect that there could be *other natural factors* of *at least as much importance* as greenhouse gases (GHGs)? Can we ask to understand them *before* committing ourselves with a *high degree of confidence*? Can we doubt and acknowledge that all natural influences on the climate are not quite known before being so confident? Is it climate skepticism or denying climate change to be curious and to want to ask these types of questions?

Heat waves – Too much alarmism, maybe?

Coming back to Koonin, a startling statement greets us on page 1 of his book *Unsettled.*

> " ...heat waves in the US are now *no more common* than they were in 1900, and that the warmest temperatures in the US have not risen in the past fifty years."[24]

Further in the book, he dedicates a whole chapter titled *Hyping the Heat* to show how he investigated the methodology used in the *Climate Science Special Report*[25] (CSSR) of the US.

I'll spare you the minutia of his investigative work but with the help of a professor at the University of Alabama who has privileged access to a meteorological dataset, they successfully demonstrate that the methods used in the CSSR are misleading. Instead of the CSSR statement that "daily heat records are more frequent than in the past," he instead proposes the following (with *great confidence*, he says):

> "There have been some changes in temperature extremes across the contiguous United States. The annual number of high temperature records set shows no significant trend over the past century nor over the past forty years, but the annual number of record cold nights has declined since 1895, somewhat more rapidly in the past thirty years."[26]

In other words, meteorologists who announce that there are *more heat records* are wrong (at least, in the United States). There is no upward trend in heat records. On the other hand, cold records are fewer. Now why is it important to produce the correct information?

Because it makes for a fairer and less alarming story than the one where torrid summers are becoming more and more common. And because it does not risk damaging the credibility of IPCC scientists. Does such a nuance undermine the thesis of global warming? No, he says.

> "As it happens, the evidence of a rise in the coldest temperatures is perfectly consistent with a warming globe – just not a "roasting" one that lends itself to graphics of bursting thermometers."[27]

Here's an example of rigor that calms alarmism and offers more accurate information. In short, what the IPCC, Pielke and Koonin all agree on is that there will be fewer extreme cold events.

"Good news!" Lomborg would tell you. Because the reality is that more people on this planet are dying of cold than heat. And by an incredible proportion! Did you know that? I didn't. I listen to the news…

In *False Alarm*, he cites a study published in 2015 in *The Lancet.* I did not believe it, so I went to see for myself the most recent study that covers 750 locations worldwide in 43 countries for the period 2000-2019.[28] The results?

- Globally, more than 5 million deaths were associated with non-optimal temperatures per year.
- **90.4% were cold-related deaths.**
- Among cold-related deaths, more than half were in Asia, almost a quarter were in Africa.

I could not believe it. Of the media's disproportionate focus on the effects of summer heat compared to cold temperatures elsewhere in the world. Are we concerned about the fate of the entire planet or not? Why not report on the climate conditions of all the regions of the world? Warming and fewer very cold nights may not be such a bad thing for these countries that often lack energy for heating purposes.

During a presentation he gave in Paris in March 2023, at the invitation of the *climato-realistes.fr* organization, Koonin also showed how some of the IPCC statements, without being false, can be incomplete and biased. He cites, for example, statement A.2.5 of the most recent IPCC synthesis report[29], "*In all regions increases in extreme heat events have resulted in human mortality and morbidity (very high confidence).*" Factually incomplete statement, he says. It should have been supplemented with information on the extreme cold that is decreasing. Another perfect example of a statement whose purpose is to persuade and not to inform.

And then, on the subject of 'non-optimal' temperatures, Lomborg would also add that alarmist projections do not take sufficient account of societies' capacity to adapt. He cites studies showing that air conditioning helped New York reduce deaths from heat waves by 66% from 1960 to 1990. Adaptation policies are essential to mitigate the effects of warming, which is not new but often overlooked due to the obsessive importance and priority given to reducing CO_2.

Of course, we North Americans and Europeans have access to reliable energy, in Quebec's case, mostly clean and inexpensive. But what about Asian and African countries? To what extent does the United Nations encourage energy supply in developing countries to counter extreme cold among other necessities?

Closing in on alarmist heat waves, why not point to an important positive effect of the atmosphere's warming? In *False Alarm*, Lomborg cites a NASA study that shows the greening of the planet. By consulting their website, we can indeed find a study *Greening of the earth mitigates surface warming*. Not only does this study acknowledge that the planet has been getting greener in recent decades, but it also shows that this phenomenon has a refreshing effect!

> "This greening-induced cooling effect was twenty-five times stronger than the warming effect caused by tropical deforestation." [30]

Isn't this great news? Instead of having an almost suicidal image of a planet drying up and boiling over, isn't it less distressing to see that in several places, it is turning green?

Back on the NASA website, an article entitled *Global Green Up Slows Warming*[31] reviews more than 250 scientific publications. The following image does not illustrate the state of the greenery or 'greenness' but how this greenness has changed, that is to say, has increased, or has decreased (brown) during the period 2000-2018. Since the image here will be in black and white, I invite you to check it out for yourself. It is very impressive.

The authors point out how obvious these changes are in places like China and India, where agriculture has intensified and where significant forest conservation and expansion efforts have taken place.

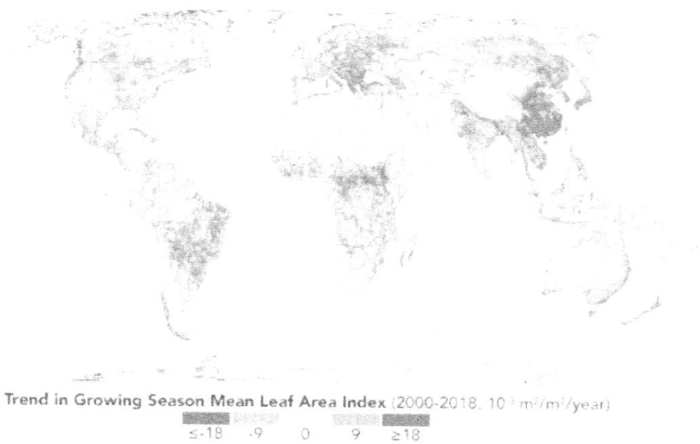

Trend in Growing Season Mean Leaf Area Index (2000-2018)
"Global Green Up Slows Warming," NASA Earth Observatory, February 2020

Pretty impressive that such greening has happened in such a short space of time, barely twenty years.

In short, when it comes to weather disasters and climate change, the United Nations' press releases, climate activists, governments, the media themselves and even some scientists, do not take the time to do the research and inform people correctly. This is not useful work.

The IPCC on cyclones, hurricanes and other whirlwinds

Admit that you are surprised that there is neither detection of a change, nor attribution to human activity for the meteorological events cited below.

Event	**Detection**	**Attribution**
Tropical cyclones	**No**	**No**
Extratropical cyclones, hurricanes, tornadoes	**No**	**No**
Winter storms	**No**	**No**
Hail	**No**	**No**
Extreme winds	**No**	**No**

And yet, this is what the extreme weather events specialist Roger Pielke Jr.[32] understands from reading the latest report. Here is a quote on tropical cyclones from the IPCC report.

> "There is **low confidence in most reported long-term (multi-decadal to centennial) trends in TC frequency- or intensity-based metrics** due to changes in the technology used to collect the best-track data."[33]

Regarding tornadoes, hail, extreme winds:

> "Thunderstorms, tornadoes, hail, lightning: "**observational trends** in tornadoes, hail, and lightning associated with severe convective storms **are not robustly detected** due to insufficient coverage of the long-term observations."[34]

Notice that we are talking about trends that can extend over decades or even centuries. In *Unsettled*, Koonin also includes statements from the IPCC Assessment Report (AR5) on these topics:

> " ...***low confidence*** in trends in small-scale severe weather phenomena such as hail and thunderstorms..."
>
> "... **confidence** in large scale changes in the intensity of extreme extratropical cyclones [storms] SINCE 1900 **is** ***low.***"[35]

He even devotes a whole chapter to hurricanes and tornadoes. "In this chapter, I intend to uncover the facts, to reveal the truth and to demonstrate that the evolution of hurricanes and tornadoes is in no way attributable to human influences,"[36] he says.

Let us assume that Pielke and Koonin understand the IPCC statements correctly: there is no detection of a long-term trend in the frequency and intensity of tropical cyclones. Nor for storms, hurricanes, tornadoes and hail. Is this reassuring? Certainly not if we refer to the media coverage of such events.

- "Climate Change – More and more intense storms to come" (*La Presse, September 25, 2022, following hurricane Fiona's passage in* Îles-de-la-Madeleine, Quebec)
- "Hurricane intensity will increase, experts warn." (*La Presse,* September 4, 2019)
- "A year of disasters fueled by climate change" (*La Presse,* December 17, 2021)

I have chosen Montreal newspaper *La Presse* headlines, but such headlines are common throughout the world's press. All you must do is search using "disaster" or "climate catastrophe" and you will be well served. The most alarming from the *New York Times* and *Agence France-Presse* will be routinely reprinted in *La Presse.*

The thing is, even according to the IPCC itself, these weather events, although they can cause huge ecological, human and economic disasters, cannot at this time be associated with climate change nor to global warming due to CO_2 emissions.

> *Remember that climate is not weather. And that rising costs of extreme weather disasters are not reliable indicators of the frequency or the intensity of such events.*

Moreover, that climate science is still very young when it comes to determining the natural influences affecting these phenomena. Could other natural influences, for example explain the frequency and intensity of hurricanes? Some new data that could tell us more and enrich climate science?

Well, it would seem that the El Niño and La Niña phenomena would have something to do with it. One particular study[37] of hurricanes in the United States found that there were almost twice as many hurricanes in La Niña years as in El Niño years. These phenomena are cyclical and occur at irregular intervals, every 2 to 7 years.

There is another phenomenon called "AMO" for Atlantic Multidecadal Oscillation. This phenomenon consists of cyclical changes in surface temperature in the North Atlantic oceans. These are 60-80 year cycles alternating between cooler and warmer phases, for over 1,000 years. An interesting passage here:

> "It is associated with changes in the frequency of North American droughts and is reflected in the frequency of

> severe Atlantic hurricanes. It alternately obscures and exaggerates the global increase in temperatures due to human-induced global warming." [38]

The graph below illustrates the relationship between the AMO oscillation and cyclones. The black line indicates the variation in the number of cyclones. Interesting, right? It clearly shows the role of oceanic movements on cyclones. And for those interested, since the early '80s we've been in a phase of warmer ocean surface temperatures in the North Atlantic, which means more cyclones.

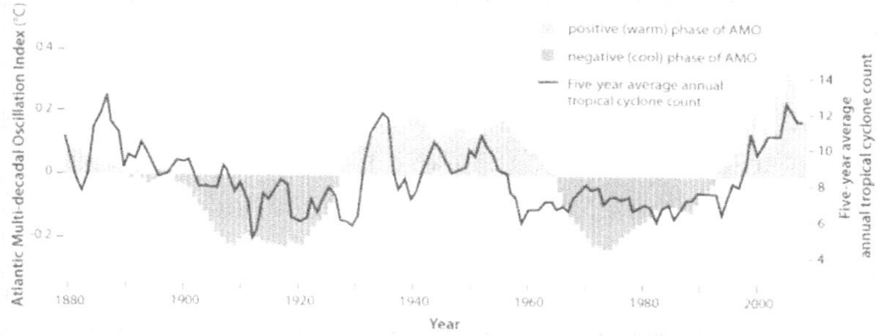

A quote from Koonin sums up the climate's complexity but more so the complexity in identifying changes specifically due to human activity.

> " ... there could be (and almost certainly are) other natural cyclic variations occurring over even longer periods.
>
> Cycles like these influence global and regional climates and are superimposed upon any trends due to human or natural forcings like greenhouse gas emissions or volcanic aerosols. They make it difficult to determine which observed changes in the climate are due to human influences and which are natural."[39]

In short, research is intense on cyclone extreme events. In fact, Koonin ends up mentioning a 2019 landmark paper[40] co-authored by 11 tropical cyclone experts. A strong majority of them were of the opinion that any change in tropical cyclones could not be attributed to anything other than natural variability (and therefore, not to human activities).

Incidentally exactly what Pielke reminds us of by quoting the IPCC itself on Atlantic hurricanes…

> "However, there is still no consensus on the relative magnitude of human and natural influences on past changes in Atlantic hurricane activity, and particularly on which factor has dominated the observed increase (Ting et al., 2015) and it remains uncertain whether past changes in Atlantic TC activity are outside the range of natural variability."[41]

… and by showing us this graph illustrating the variation (or rather absence of trends) in the total number of cyclones occurring on the globe. The bottom line indicates "major" cyclones, which are Class 3, 4 or 5 cyclones that reach over 180 km/h (or 96 knots). Clearly, over this relatively short 40-year period, there are years when there are more major cyclones and there are years when there are fewer, but there is no upward or downward trend for the entire period.

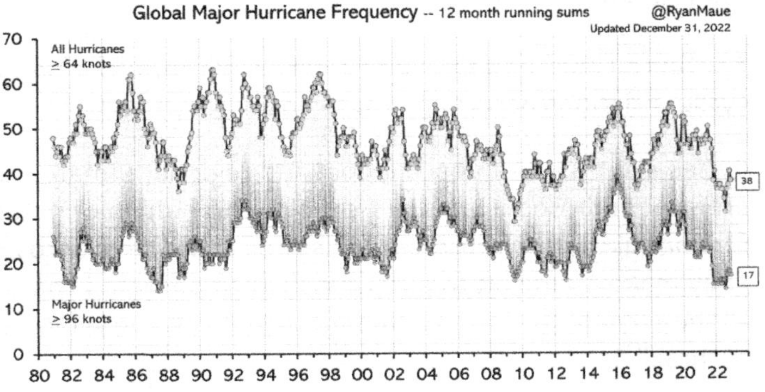

Source : Dr. Ryan N. Maue –
Global Tropical Cyclone Activity
https://climatlas.com/tropical/

Blatant disinformation in the IPCC regarding cyclones

You must be a passionate nerd to dissect the IPCC assessment reports. At the very least, you need to have a very detailed knowledge of the subjects covered. Fortunately, there are such experts such as those I cover in this book. Generally, they find that the IPCC Working Group 1 (WG1) provides good quality information. As for the other groups (WG2 and WG3), and particularly the *IPCC Summaries for Policymakers*, their work deserves to be scrutinized since it is not uncommon to find important errors.

Later, a full chapter deals with the IPCC, models, and scenarios, but as an appetizer, here are some findings on the subject of cyclones.

The first case concerns the *cherry-picking* of cyclone studies by the IPCC, whose duty it is to carry out a systematic review of *all* scientific literature. Remember reading earlier that when assessing on cyclones, the IPCC referred to "*changes in technology used to collect the best-track data*" to justify its statement on detecting changes in tropical cyclones? In other words, no change was detected but it was due to the use of new data collection instruments?

The thing is, the IPCC is admitting here having replaced a "dataset of the highest quality available on tropical cyclones around the world and widely used in research (the *International Best Track Archive for Climate Stewardship (IBTrACS))* with some "new technology."[42]

So, in August 2021, when Roger Pielke reports on the IPCC's findings, he finds bizarre their denigration of and doubts on the dataset's reliability. After all, his research career began with a study of cyclones (this guy knows about cyclones ...) and the discovery of a link between their frequency and La Niña. At the time, he noted that:

> "The denigration of the TC (tropical cyclone) 'best track' dataset is bizarre. The dataset is the highest quality available on tropical cyclones around the world and widely used in research. It'd be a shame if the IPCC process were to have been used to promote certain work by denigrating the widely recognized best available data." [43]

Almost two years later, after a detailed excavation of the IPCC report, he published a new text *Misinformation in the IPCC*[44] in which he revisits the IPCC's "new technology" used for the assessment on cyclones. What he found was a statement that suggested that *normalized hurricane damage increase* in the United States *can* be used to contradict data judged by scientists to be the "*best-track dataset.*"

However, as he points out, *one cannot use economic damage data* to infer climate trends, let alone use them to question or doubt the validity of actual climate data.

Worse, it seems that in this case, the IPCC authors chose and cited one study in particular, while the other eight, which went against it, were not. Strange when we expect the IPCC "to review the scientific literature. *All of it* – that means including more than just a subset of studies that can facilitate the construction of a narrative to the liking of the authors," he said. "You won't find a more obvious example of cherry-picking of the literature by the IPCC," he concluded.

A second case shows the misinformation that can result from inconsistencies between different levels of IPCC reports. If the *Sixth Assessment Report (AR6) of WG1* stated that there was no detectable trend for tropical cyclones (TC),

> "There is **low confidence** in most reported long-term (multidecadal to centennial) trends in TC frequency – or intensity – based metrics."

the *Summary for Policymakers* states that changes have been observed since the last Assessment Report (AR5) and that it could even be attributed to human activity!

> "**Evidence** of observed changes in extremes such as heatwaves, heavy precipitation, droughts, and tropical cyclones, and, in particular, their attribution to human influence, **has strengthened** since AR5."

To make a long story short, Pielke's findings reveal:

- That the study selected and cited by the *IPCC Summary for Policymakers* did not focus on the number of cyclones but on the number of "observations" of cyclones (a cyclone can be "observed" in several instances).
- That the authors of the study cautioned that the detection of change, and therefore attribution to anthropogenic factors, could not be inferred.
- That a major amendment had been published and sent to the IPCC within the timeframe that would have allowed the IPCC to take it into account. In fact, the correction specified that the findings of the study applied to only 2 of the 7 cyclonic basins (where cyclones form on the globe), and therefore accounted for less than 20% of all global cyclonic activity.

But having found the source of the error, Pielke is still not content. Digging deeper, he finds another error that he considers unforgivable. Not only do the Summary reports for policymakers go into fiction mode by falsely claiming that there is detection of cyclonic activity change, but they also state that the trend has increased since the last report in 2014. What's really happening? Are there changes or not? Is there a higher proportion of *force majeure* cyclones (96 knots or more) compared to less severe cyclones (64 to 96 knots)? If you look at the chart that follows, it looks like there *is* a trend, but notice the time interval 1980-2022.

"Trends in the Proportion of Major Hurricanes." 1980-2022, *Substack "The Honest Broker,"*, April 2, 2023

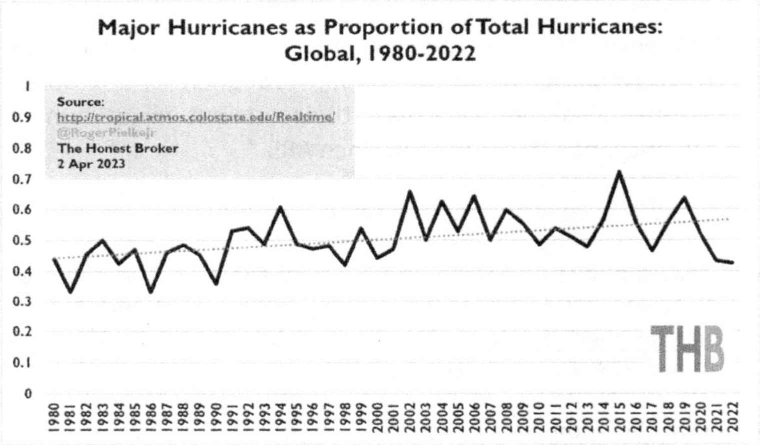

But then, going back in time to cover the period 1950-2022 (instead of 1980-2022), here is what we get: no trends at all!

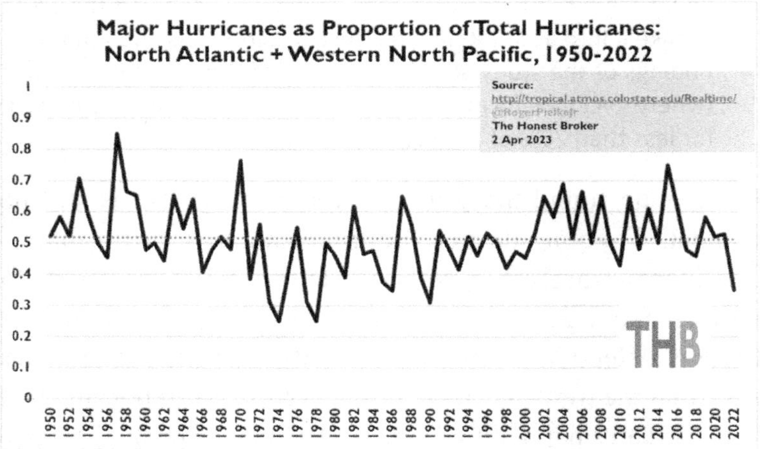

Okay, but where does that lead us?, Pielke asks. What should we understand? Well, we should understand that it is possible to arbitrarily choose a dataset to confirm any preconceived idea. Intentionally or not, there is misinformation.

- You want to show an upward trend? Begin your analysis in 1980.
- You want to show that there is no long-term trend? Start in 1950.
- You want to show a downward trend? Start in 2002.

After all this, it's hard not to conclude that the IPCC reports, at least the Executive summary reports, can be misleading. Not only their authors may be tempted to favor some studies while neglecting others to validate some narrative, the time intervals chosen can also be misleading.

The IPCC on flooding and drought

In November 2021, the Fraser Valley of British Columbia in Canada was flooded. Major roads were submerged or washed away, and landslides blocked portions of roads. The images presented by *Radio-Canada*[45] were striking and heartbreaking. Thousands of people evacuated, farms, animals and cultivated fields flooded, telephone lines cut off from the world.

"Definitely an extreme event," according to Brent Ward, Professor of Earth Sciences at Simon Fraser[46] University, "… and we believe these could be the worst landslides and floods since perhaps 1983".

Maybe, but if you do a quick search on the beautiful Fraser River, you will see that the region has seen much worse. You will be fascinated by images of the 1894[47] floods (a time when no major event could be attributed to CO_2). As the river rose day after day, residents only realized the severity of the event when trains were interrupted. A 25-feet rise of the river, can you imagine? What an era! Today we would say that such events are caused by climate change…

So, what does the IPCC Assessment Report (AR6) have to say about the detection of high river flows?

> "In summary there is **low confidence** in the human influence on the changes in high river flows on the global scale."[48]

…and on attribution to climate change due to human activity?

> "In general, there is **low confidence** in attributing changes in the probability or magnitude of flood events to human influence because of a limited number of studies, differences in the results of these studies and large modeling uncertainties."

In the 2015 assessment report (AR5), Koonin also noted the very low level of confidence regarding the increase in global flood intensity and/or frequency. Moreover, some floods are no greater than those occurring in the last 500 years. Koonin recalls an excerpt from AR5:

> "In summary, there is *high confidence* that past floods larger than those recorded since the 20th century have occurred during the past 500 years [in some regions]."[49]

In other words, he says, there has been worse flooding in the last 500 years. But did we not see at the beginning of this chapter the detection and attribution of heavy precipitation to climate change?

> "In summary, the frequency and intensity of heavy precipitation have likely increased at the global scale over a majority of land regions with good observational coverage."[50]

And doesn't more rainfall cause more flooding? Again, not necessarily, according to the IPCC. No link between heavy rainfall and increased flooding. Totally counterintuitive … But that is what the IPCC "science" says.

> "However, heavier rainfall does not always lead to greater flooding."[51]

OK, now this is where it gets interesting. Bold characters are mine.

> "Flooding is also affected by **changes in the management of the land and river systems**. For example, **clearing forests** for agriculture or **building cities** can make rainwater flow more rapidly into rivers or low-lying areas. On the other hand, **increased extraction of water from rivers** can reduce water levels and the likelihood of flooding."[52]

To me, this is very interesting because the IPCC introduces 'human' factors other than CO_2 emission, factors which can cause floods or even dry up rivers. Take the Colorado River, for example. A fascinating case, reported here by Agence France-Presse (AFP), May 2023:

> "The river supplies about 40 million people with water and irrigates millions of hectares of arable land to feed America. But in recent years, its flow had fallen by about a third: its level was so low that the American West risked disaster."[53]

The Colorado River is "threatened by over 20 years of drought *worsened by global warming*," AFP adds. By global warming? Not really. For the southwestern states of the United States, drought has been a problem for millennia, says

Steven Koonin in *Unsettled.* What follows is an image taken from the IPCC AR5 Assessment Report. For information purposes, the PDSI can range from -10 (very dry) to +10 (very wet), but most values are between -4 and +4.

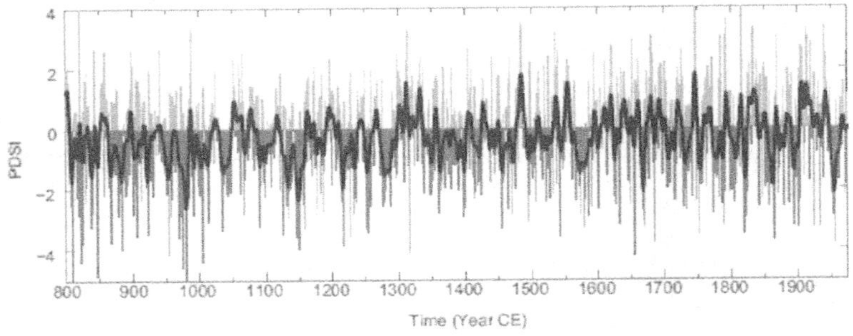

FIGURE 7.9 **Annual values of the Palmer Drought Severity Index for the Southwest US.** The heavy black line is smoothed behavior over nine-year intervals. (AR5 WGI Figure 5.13.)[19]

Source: Koonin's *Unsettled, p. 139*

"This chart retraces droughts in the southwestern states over a period of more than 1,200 years," Koonin writes. We can see that there were many periods of drought over several decades.

> "Any droughts before 1900 cannot have been due to human influence, and the so-called megadroughts from 900 and 1300 AD are thought to be associated with the naturally warmer globe at that time." [54]

As for the Colorado River, let's see what we can learn from a reconstruction of the river flow from the 700's to the 2000's.

Note the long droughts around the years 900, 1115 and 1600 that affected the river's flow.

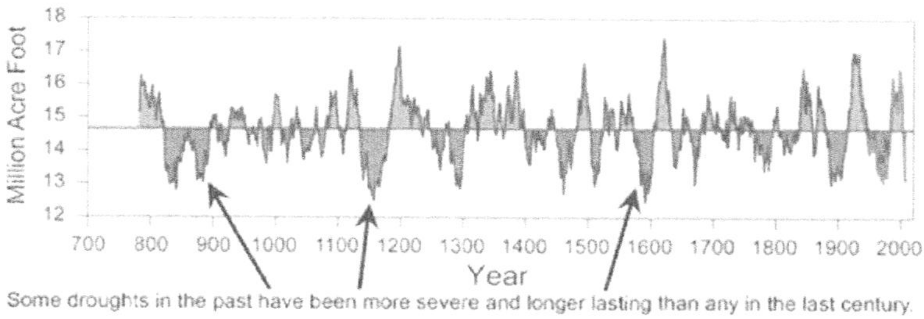

FIGURE 7.10 **Colorado River flow over 1,200 years, reconstructed primarily from tree ring analyses.** (Figure from NCA2009.)[24]

Source: Koonin's *Unsettled, p. 140*

But back to the floods in British Columbia. In a fine example of good journalistic work, *La Presse* recalls that "a huge lake was located a hundred years ago in the agricultural plain of Sumas," a region affected by the floods of November 2021. Then, in the 1920s:

> "A floodway channel, dykes and a pumping station were built in the early 1920s, transforming this vast body of water into a fertile plain, which now feeds much of the Canadian province of British Columbia with its many dairy and vegetable farms as well as its egg and poultry production."[55]

An example of human intervention in the 1920s with occasional side effects? It is certainly a good question. But again, nothing to do with CO_2.

Back to the present. Other times, other priorities: Montreal's bicycle paths, for example. They are a source of pride for the mayor, but when there is heavy rain, it's not unusual for the basic infrastructure to fall short. "It's climate change!" says the mayor.

What if some flooding resulted from poor maintenance of municipal water and wastewater infrastructure? Why aren't we asking these questions? What if the different levels of governance had erred when it comes to prioritizing safety and public health emergencies? What if more adaptation policies were developed before spending millions and even billions on zero emissions (or on bicycle paths)? (Okay. I'll stop here...).

Another example. Baie St-Paul, May 2023. The Rivière du Gouffre in this magnificent Charlevoix region bursts its banks and floods an entire district of the town.[56] A moving testimony of a resident of Baie St-Paul is published in Montreal's *La Presse*. With a master's degree in geography, Philippe Juneau calls on political institutions to review their regulations.

> "Unprecedented floods? Catastrophic? Phenomenal? That's not what science tells us."[57]

He then goes on recalling data from the *Centre d'expertise hydrique du Québec*, which monitored the Rivière du Gouffre. In 1968, 1973, 1976, the river produced even greater flows than the floods of May 2023.

> "Both deaths are the result of political inaction [...]. The residence where the tragedy occurred is in a flood zone, literally less than 100 m from the inner bank of a bend in the Gouffre River, the first area to be flooded by the stream."

Where are we with adaptation measures in Baie-St-Paul? Well, it seems that there was indeed a project submitted to the Ministry of Municipal Affairs and Housing. Unfortunately, we learn from journalist David Rémillard of *Radio-Canada*, that the request had been refused. Baie-St-Paul will therefore live for an indefinite period, the effects of a river that roars periodically.

> *In conclusion, instead of raving about climate change as the cause of all evil, perhaps it would be better to question our adaptation policies, to care for nature's elements and to protect wildlife and people who its greatest beneficiaries are.*

The IPCC on wildfires

Hurricanes, floods, and forest fires produce images that are heartbreaking. We would like to think we can control these forces of nature, have an emergency push button that calms storms, but we soon recognize our smallness and our powerlessness in the face of these extreme phenomena.

Are they more frequent, more intense due to climate change? In the case of cyclones and floods, as we have seen in the previous sections, this cannot be confirmed based on IPCC assessments. What about forest fires?

First, as surprising as it may seem, the IPCC does not assess wildfire trends. Instead, the IPCC speaks of "weather conditions that promote wildfires (fire weather)." On his *Substack*, Roger Pielke summarizes what the IPCC means by fire weather.

> "Weather conditions conducive to triggering and sustaining wildfires, usually based on a set of indicators and combinations of indicators including temperature, soil moisture, humidity and wind."

In fact, according to the IPCC, the weather is not the most important factor in fire: "human activities have become the dominant driver. Indeed, most wildfires are started by human activity."[58]

That said, the IPCC expresses a medium level of confidence (50-50) that there are positive trends, in the last century, in fire-prone conditions (or fire weather) *in some regions*, including southern Europe, northern Eurasia, the United States and Australia.

Simply put, according to Pielke, the IPCC does not claim to attribute the conditions conducive to fires (fire weather) to climate change. In addition, it is silent on any trend of fire numbers and area burned. Surprising, if we compare these IPCC statements with those of the eminent climatologist Michael Mann (remember the creator of the hockey stick?).

Can these fires [Canadian forest fires in summer 2023] be attributed to climate change?" asks an MSNBC reporter.

> "This is a sign of **far worse things to come** if we don't rein in fossil fuel burning and carbon emissions and continue to warm the planet.
>
> **The only way** to prevent these events from becoming more frequent and more intense is to prevent the continued warming of the planet. And the **only way to do that** is to decarbonize our economy as rapidly as possible."[59]

I don't know about you, but when there is only one magical solution to all the earth's problems, it seems to me a bit simplistic. I remember an article in *Le Quotidien* telling the story of *La petite histoire des grands feux du Saguenay-Lac-St-Jean.*[60] A well-written review of the region's forest fires, especially the great fire of 1870. Heat wave, farmers who burned accumulated wood after land clearing, west wind that rises and finally, the lightning that strikes on the dried-up forest by the heat wave. It's all there. Perfect fire weather conditions.

"In total, 380,000 hectares were destroyed, three and a half times the surface area of Lac Saint-Jean. The fire left seven dead and hundreds injured. Five hundred families lost everything they had." The article also recalls the major fires of 1846, 1920 (83 fires), the late 1940s, 1996 (378 forest fires in operation in Quebec), and 2005, which had a record of 1374 fires.

The part of the article I found interesting are the testimonies of some of the citizens of 1870. Oddly enough, they very much resembled today's climate activists' who speak of the immorality of using fossil fuels. A citizen of 1870:

> "One would have to be blind not to see in these various scourges descending on us simultaneously, a serious godsent warning. We have become selfish, proud, cowardly."

As for Michael Mann who claims that forest fires are mainly due to CO_2, "this is obviously false", retorts specialist of extreme weather events, Roger Pielke. "There are very good reasons to decarbonize the global economy, but doing so to control forest fires and their impacts is not one of them."[61]

And then, what does the data say? California, a victim of forest fires, if there is one, surely suffers from the effects of climate change, right? In *Apocalypse Never*, Shellenberger asks the question to "fire ecologist" Dr. Jon Keeley. He studies the historical patterns of fires and their relationship with climate and urban development.

> "We've looked at the history of climate and fire throughout the whole state, and through much of the state, particularly the western half of the state, and we don't see any relationship between past climates and the amount of area burned in any given year."[62]

In *False Alarm,* Bjorn Lomborg takes note of the gap between the alarmist narrative and the evidence of the past 150 years. A study covering six continents and two millennia shows that, in fact, the fire trend has been *decreasing* since 1870[63]. Similarly, here is Koonin in *Unsettled*:

> "Sophisticated satellite sensors first began monitoring wildfires globally in 1998. Unexpectedly, analysis of the images showed that the area burned annually declined by about 25 percent from 1998 to 2015."[64]

Interestingly enough, NASA[65] has recently been tracking the total area, worldwide, of devastation by fires every year. Koonin here takes a graphic directly from their website. Here are the standard values of areas destroyed each month by fires and the (quite clear) trend.

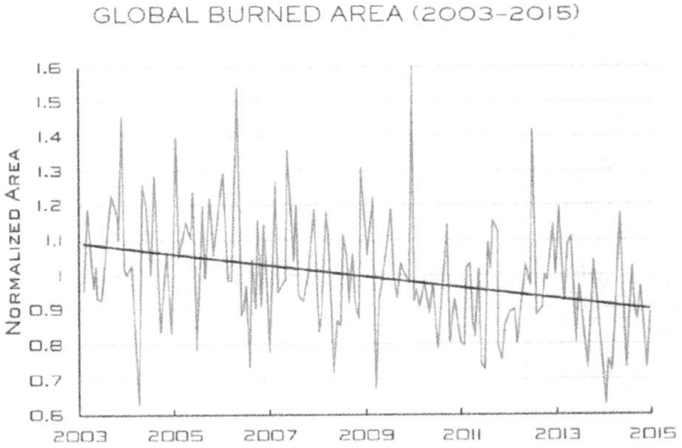

FIGURE 7.12 **Normalized values of the global area burned by fires each month (gray line) together with their trend.**[32]

Source : Koonin's *Unsettled, p. 143*

Hold on! More importantly, this *decline*, according to some researchers, is attributable to human activities, in particular the expansion and intensification of agriculture. So yes, *there would be* human activity with positive effects! For example, land clearing machines replaced brush clearing with fire.[66] Isn't this good news?

Speaking of human influences other than CO_2 emissions, forest management, fire management when they occur (do we let everything burn or do we intervene?), the presence of buildings near vulnerable areas and human behaviour *can* mitigate the devastating effects of fires.

June 2023, forest fires ignite several regions of Canada. Once again, alarmists on one side and so-called deniers of global warming on the other compete for media presence, ignoring the honest citizen who would like to learn more on the subject. "Climate denial is gaining ground"[67] reports CBC. One wonders what media researchers are paid for... Evening televised panels on conspiracy theorists. "Are we likely to repeat the social climate we experienced during the pandemic?" asks a TV host. Very disappointing… I must say, total failure of the mainstream media.

Anyone wanting to know more had to go on official websites, consult studies and even the American media to have an idea of what was going on in our own region... For my part, a Twitter account @IWeatherON offered me an explanatory and totally fascinating video.[68] The account holder said he had received several questions about the simultaneous outbreak of fires in northern Quebec. He was also curious about it. Using satellite data, the video superimposes fire hotspots on the lightning bursts, followed by very powerful winds that ignite the whole thing. That was it. No conspiracy here. Move along.

On several other Twitter accounts, charts taken from the *National (Canadian) Forest Database* illustrated the *downward trend* in fires across Canada, and here in Quebec, over the past three decades.[69]

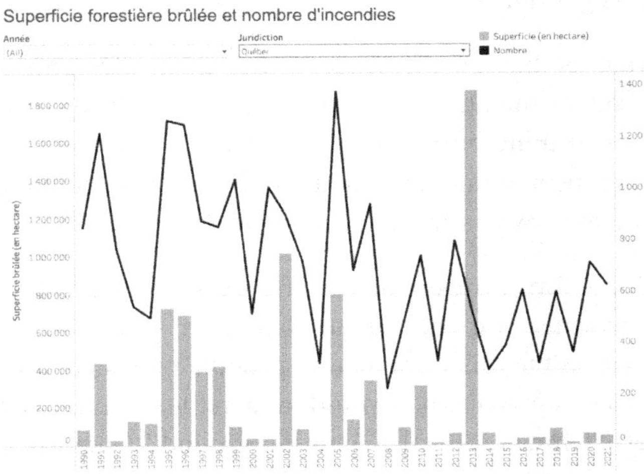

Others present graphs showing the percentage of forest fires in Quebec caused by human activity (negligence), either in number or in areas burned. Hint: These percentages are on the rise.

Let's understand each other. Extreme weather events cannot only destroy environments but take a toll on human lives and the wildlife within them. But in my opinion, when this happens, simplistic solutions as stated above are not enough. The media have a responsibility to educate and contextualize any event in a historical setting instead of lapsing into juvenile alarmism.

Fortunately, we can manage to find well thought-out texts that inform us correctly. One such text is signed by three forest fire and forest ecology specialists Victor Danneyrolles, Raphaël Chavardès and Yves Bergeron. Titled *North American boreal forests burn a lot, but less than 150 years ago*, they explain the dark sides, but also the benefits of forest fires. Then they inform us of something that may be surprising for some. Remember. The official CO_2 alarmist narrative postulates that "climate change" is essentially due to the emission of CO_2 starting around the 1850s.

> "We gathered 16 studies that had independently applied the same method to different areas across North American boreal forests, from Alaska to Québec.
>
> After reanalysis of all this data in what scientists call a 'meta-analysis,' the results[70] are striking: North American boreal forests burned much more 150 years ago than they do today. In the earliest period covered by our data, between 1700 and 1850, the annual area burned was between two and more than 10 times greater than what has been observed over the past 40 years." [71]

Makes you wonder, no?

The IPCC on sea levels

Scenarios of swallowed up vanishing cities and islands submerged by the seas abound in the official media and magazines dedicated to apocalyptic activism. Strong images of New York's Statue of Liberty[72] that bathes in water up to its waist and continents whose edges are found below sea level are gripping and troubling.

There's no doubt about it. In some areas, sea levels are rising. Remember Koonin's *Unsettled* chart presented in chapter 2?

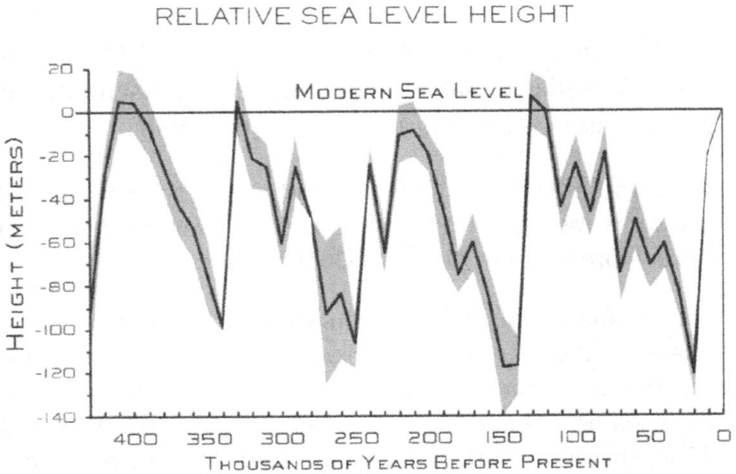

FIGURE 8.1 **Global sea level over the past 400,000 years as estimated from geological proxies.** The typical uncertainty in these estimates is 10 meters.[2]

It shows that every 100,000 years, sea levels *slowly* drop by about 120 metres (about 400 feet), and then rise *much more quickly* for 20,000 years or so. At this time, glaciers melt and sea levels rise.

Where are we now in this sequence of glacial and interglacial periods? As we already mentioned, in an interglacial period called "Holocene". And according to geological observations, the seas *did* go up "about 120 metres (400 feet) starting 20,000 years ago to about 7,000 years ago, when rising waters slowed dramatically"[73] And so, on a time scale of several thousand years, at this time, the rising waters have slowed down.

On a shorter timescale, however, this rise in sea level is not surprising according to Judith Curry. After all, during the *Little Ice Age*[74] (1450-1850), glacial mass had increased, and so, it is normal that sea level rise resumes when small warmings occur.

But the IPCC goes further in its analysis by looking at the *speed of the observed increases*. Sea level rise is accelerating, especially in recent decades, they find. It is rising, but faster and faster! It goes from 1.3 mm/year (1901-1971) to 1.9 mm/year (1971-2006) to 3.7 mm/year (2006-2018).

Of course, this gives us headlines such as these:

- "Rising sea levels, a time-delayed disaster." - *Radio-Canada Info*, taken from Agence France Presse, October 12, 2021
- "With rising ocean levels, the UN fears an exodus of biblical proportions." *Le Devoir*, Agence France-Presse at the United Nations, 14 February 2023

We get the message, but should we really be concerned about such accelerations in sea level rise? Perhaps not as much as they say we should, Koonin tells us. The speed at which the sea level rises varies across decades. Take the period 1920-1950, for example, he says. "Sea level rise [during these decades] was almost as rapid as what recent satellite data tells us, about 3 mm/year."[75]

Another example comes from the *United States Climate Science Special Report*, which reported that "Global mean sea level (GMSL) has risen by about 7-8 inches (about 16-21 cm) since 1900, with about 3 of those inches (about 7 cm) occurring since 1993". But, says Koonin,

> "The fact that three of the seven inches of rise since the dawn of the twentieth century occurred in the past twenty-five years does indeed seem alarming – but suddenly less so if you know that it also rose almost as much (6 cm, as opposed to 7 cm) in the twenty-five years between 1935 and 1960." [76]

Can you see here what the scientist is doing? And how important it is to resist alarmism? If you go back in time, you'll often find similar phenomena, or ones that repeat themselves on a cyclical basis.

To further illustrate his point, Koonin shows, as an example, the monthly mean sea level anomaly at the tip of Manhattan. This is a place where the speed of sea level rise has been measured for 160 years. The horizontal line is the average speed of sea rise at this point, around 3 mm/year.

The values above this line indicate that the rise is accelerating; the values below show that the rise is continuing but not as fast as the average speed of 3 mm/year. Let's be clear: it rises all the time. We're talking about the variations in the *speed of the rise* here.

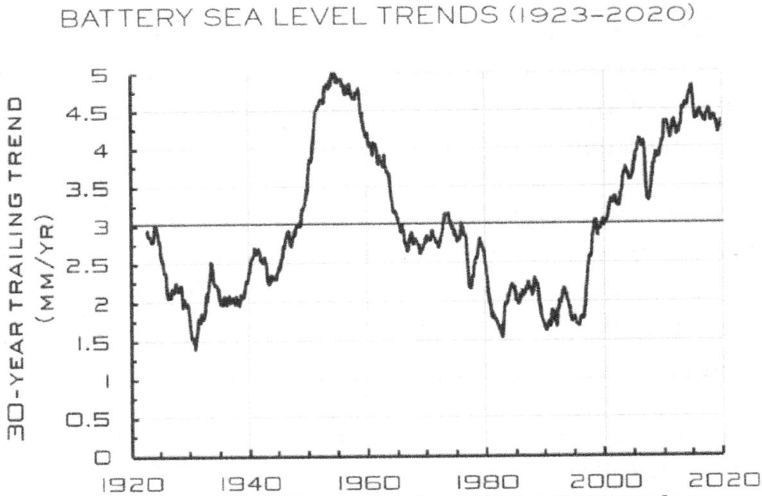

FIGURE 8.8 **Thirty-year trailing trends in the sea level at The Battery from 1923 to 2020.** Uncertainties (1σ) are about 0.35 mm/yr. The horizontal line indicates the average rate of 3.02 mm (0.12 inches) per year.

Source: Koonin's *Unsettled, p. 164*

Are you seeing what he's seeing? I admit it took me a while. There is a cycle here of about 60 years. Moreover, it is in sync with the Multidecadal oscillation of the Atlantic (AMO) that we saw earlier. Indeed, many researchers recognize that sea levels can be associated with natural cyclical phenomena lasting 60 years.

But then again. Why worry and look in the past for periods that reproduce the same phenomena as those of recent decades? Why worry about cyclical phenomena at all? Icebergs are melting! The sea is rising! That is a fact. That sea levels rise more or less quickly does not change anything, does it? "We will still be flooded", some will say. That this rise is influenced by cyclical phenomena, what does that change?

What changes is that if the rate of sea level rise is influenced *by other cyclical phenomena*, it may mean that it is a little less or not at all influenced by human influences and CO_2. It makes us realize that it is very important to want to understand *all the influences* on climate.

> *This is what scientists mean when they insist on the importance of distinguishing natural influences from CO_2 influences. Otherwise, global organizations and our governments risk investing in wishful thinking about "transformational change" rather than investing in adaptation policies that more realistically mitigate the impacts of climate disruption. On that, there should be consensus.*

Here, Koonin:

> "The message put forth that human-caused warming is the sole source of rising sea levels gives the impression that reducing emissions is a solution [...] Alas, even if we were the

> culprit and ceased all emissions tomorrow, global sea level would continue to rise.
>
> If sea level does become a serious threat in coming decades, there's no doubt that we'll be better prepared having devoted resources to further research and adaptation […]"[77]

Disaster scenarios, there are plenty. Curry, whose job is to assess uncertainties and risks associated with climate phenomena, considers that achieving proper assessments are enormously challenging.

The accelerated loss from the West Antarctic Ice Sheet (WAIS) is one that is very much associated with all worst-case scenarios of twenty-first century sea level rise, says Curry. "Estimates of the contribution to global sea level rise from a complete collapse of the WAIS are 3.2-4.2 meters."

> Another storyline relates to geologically-induced melting at the base of the ice-sheet – a factor not mentioned in the IPCC assessment reports. The West Antarctic Ice Sheet lies atop a major volcanic rift system with 138 documented volcanoes that are widely distributed throughout West Antarctic. These volcanoes produce a steady flux of heat, but episodic active volcanoes can potentially produce significant local melting, particularly under the vulnerable Thwaites Glacier."[78]

But "the impact of under ice volcanoes in terms of producing significant and varying under ice melting, is as of yet an undetermined effect on the glacier mass balance," she says. As is shown by a NASA scientist in an article published in May 2020 in which he explains why he believes volcanic activity does not melt polar glaciers.[79]

In other words, the search continues.

* * *

In closing this chapter, weather events are a centerpiece of the UN official narrative of the past 30 years. To demonstrate the "danger" of climate change, it's easy to understand why the IPCC's work is obsessively focused on the worsening and dangerous nature of extreme weather events.

> *But here's the thing. As seen in the last IPCC Assessment Report (AR6) and in this chapter, the detection and attribution of all extreme events to CO_2 warming is not what "the science" says.*

A sad commentary at a time when investments could be used more for adaptation policies than for radical carbon neutrality programs, which for the time being are more of a utopian fantasy.

CHAPTER 4

The IPCC, its climate models and scenarios

Climate modeling

Imagine earth. The terrestrial globe. Imagine covering it with a grid of three-dimensional boxes. Typically, we will have ten to twenty layers of these boxes, each of them covering 100 km by 100 km above the ground. For the cells covering the oceans, they will be smaller but there will be more vertical layers. This is what Koonin illustrates in the chapter named "Many Muddled Models" of *Unsettled.*[1] In the end, we will have one million boxes for the atmosphere and one hundred million boxes for the ocean. A hundred million boxes!

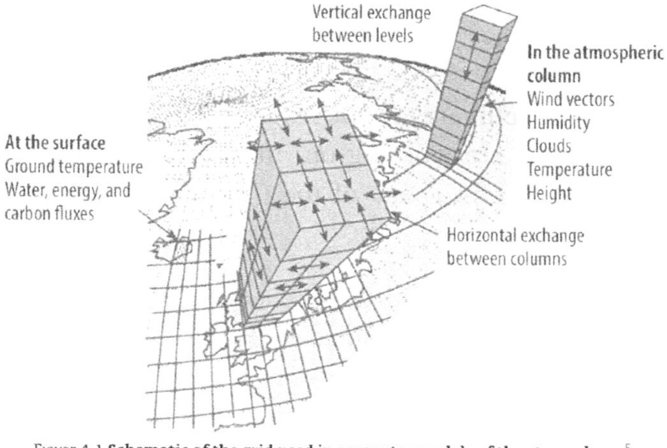

FIGURE 4.1 **Schematic of the grid used in computer models of the atmosphere.**[5]

Source: Koonin's *Unsettled, p. 164*

Now imagine that your job is to produce computer climate models that will predict the climate in the year 2100. Each of these boxes will have to simulate its inner climate (humidity, temperature, winds, atmospheric pressure, etc.), and make it evolve over time. Moreover, if you succeed, your model should be able to reproduce the climate of the last century.

Why should it? Because we *know* the temperatures and climate of the last decades. So, if the dataset is reliable, if we know the variables that influence the climate (GHGs and other natural influences), the model should illustrate that the temperature has increased by 1.1°C between 1900 and 2020 and that there have been temperature fluctuations throughout the period. For some decades, temperatures rose faster than others, but temperatures decreased between 1940 and 1980 (remember the American publications that announced a new ice age in 1975!).

After all, the model's credibility requires it to reproduce the past. "If you can't reproduce the past, be skeptical,"[2] Koonin says.

Predicting the climate of the future. This is what many modeling teams are working on. In the end, the IPCC will average results from a few of the models produced around the world.

An important part of 'climate science' is therefore based on computer programs, modules, supposed reliable data, approximations, and assumptions. And all this moves through time! Absolutely fascinating! And no wonder "model results differ dramatically both from each other and from observations"[3] of the past.

Limitations of climate modeling

For Koonin, the fact that the temperature simulated by different models varies by as much as 3°C is one "particularly jarring failure."[4] After all, *it is three times more* than the warming observed throughout the 20th century that they claim to describe and explain! Also, "The fact that the spread in their results is increasing is as good evidence as any that the science is far from settled,"[5] he writes.

And for good reason! He quotes one of the lead researchers saying "Cloud-aerosol interactions are on the bleeding edge of our comprehension of how the climate system works, and it's a challenge to model what we don't understand."[6] And one of the reasons for this is that aerosols generally cause cooling that partially counteracts warming... (emphasis is *mine*)

> "In other words, [says Koonin], we don't really understand an influence on the climate system that's *about the same size as the human-caused warming influence*."[7]

Another thing that bothers him: computer models do not reproduce the strong warming observed between 1910 and 1940.

> "More bluntly, they're saying that we've no idea what causes this failure of the models. They cannot tell us why the climate changed during those decades." [8]

And this is "deeply unsettling", he says, because the warming observed between 1910 and 1940, (at a time where human influences were not very important) is comparable to the one observed at the end of the last century. Isn't this strange? Could it be that CO_2 is less important than we think?

"We are looking at climate in a hundred years and yet we have terrible data to describe what happened in the past," in conversation with Jordan Peterson.

Climatologist Judith Curry also analyzes the different models in her book *Climate Uncertainty and Risk*. Deep disagreements exist about the accuracy that simulations of climate models can offer, she believes. The climate system is non-linear, complex, and poorly understood in theory. She gives the example of glacial plates, of the internal variability associated with ocean movements, the indirect effects of the sun, geothermal flows under the sea and under glaciers (remember that there are 138 volcanoes under vast sections of Antarctica!).[9]

Finally, to fully grasp the complexity of climate science, here is a list of uncertainties taken from a recent *Columbia Climate School* article that I found interesting.[10]

- The role of clouds that influence and can be influenced by climate
- Water vapor and aerosols that have a greenhouse effect but can also cause cooling
- The natural changes in the climate caused by volcanic activity, sun intensity variability as well as El Niño et La Niña
- Ocean circulation changes like the Atlantic Multidecadal Oscillation (AMO)
- The consequences of thawing permafrost that releases CO_2 and methane, two GHGs
- The speed at which ice sheets are melting

Let's stop there but we get the idea and especially the impossible task of parameterizing all these phenomena in computer programs. Add to that the limited processing capacity of computers and we are far from a "settled" science, which is quite normal. In fact, the article concludes with this wise remark:

> "Climate scientists are still uncertain about a number of phenomena. But it is the nature of science to never be settled — science is always a work in progress, constantly refining its ideas as new information arrives."

The IPCC's scenarios

The political importance of scenarios

This is probably the section I had the most difficulty with. The material is more technical but everything about climate scenarios deserves attention. In another life, I held positions as an IT systems architect so I had to control my curiosity and make a choice about what I would cover. I hope I have managed to present the essentials but know that this section is one of the most important ones in the book. Why?

First, because huge expenditures, government subsidies and regulations are justified based on IPCC models and scenarios. They are, rightly or wrongly, what is presented in the media and what drives GHG reduction and adaptation policies. Also, because it helps us recognize that the scenarios most used by environmental groups and the media are the most catastrophic of the lot. Whether we are talking about CO_2, energy transition, regulation, or "civilizational transformation," we must know that all this takes root in our governments' trust (belief will do…) in IPCC climate scenarios.

Because if we are to believe what circulates in the mass media, we should be inhabited by a sense of urgency before "the cataclysm becomes even more cataclysmic."[11] For sure, according to climate apocalypse activists, we are witnessing the worst-case scenario. Here. Right now!

Yet, in the latest 2021 *IPCC Assessment Report*, new climate scenarios emerged. Five were selected from hundreds and each one of them proposes a possible future although most are highly improbable (including the cataclysmic scenario). There are also assumptions of GHG emissions, for example, that simply no longer hold water.

Scenario names and the concept of "forcing"

But to understand what we are talking about, I had to learn a few new terms. First, how IPCC scenarios are named. The first part (SSP#) indicates a socio-economic development path while the following figure gives an indication of the human influence in terms of warming (e.g.: SSP2-4.5).

These socio-economic development paths took about ten years to develop. They are generally well received although they are not immune to criticism.

As for what follows "SSPx," beware! The last figure does not indicate global warming in the number of degrees compared to the reference period 1850-1900. In other words, the SSP2-4.5 scenario does not mean that the temperature in 2100 will have increased by 4.5°C since 1850.

This number, 4.5 in our example, gives an indication of the warming caused by human influences, what scientists call human caused forcing." The higher the number, the greater the "forcing" and therefore the more warming there would be.

I hear you asking, "Okay, but what does 4.5 stand for?" Well, this is where it gets interesting. For reference, Koonin states that "the net human influence today is most likely between 1.1 and 3.3 Watts/m^2,[12] while climatologist Curry talks about 2.5 Watts/m^2.[13] All right. They concur. Let's go with 2.5. Now, since the heat radiated by the earth amounts to an average of 239 Watts/m^2, it's as if human influences amount "to a little more than two light bulbs of 100 watts per square meter... so around 1%."[14] The title of the chapter in which Koonin explains all this is "Humble human influences."

We can understand why...

The most recent IPCC's scenarios

Going back to IPCC scenarios, there are five of them: two optimistic, one moderate, one pessimistic and one who is completely disconnected from reality. We also note that there are two SSP1 scenarios with different forcings. This is the image that appears in the *Summary for Policymakers* of the Working Group #1 (WG1) report.

GLOBAL SURFACE TEMPERATURE CHANGE RELATIVE TO 1850-1900*

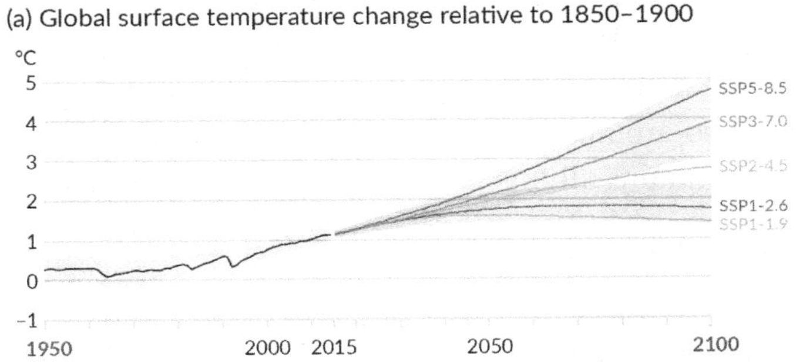

The first table that follows gives the temperature changes up until 2100 while the following table gives us an idea of what happens to CO_2** emissions.

The figure in brackets shown in the last column of the 2nd table consider that we already have +1.1°C warming since 1850.

---

* GIEC, Climate Change 2021 : The Physical Science Basis, Summary for Policy Makers, p. 22

** Id. p. 14

ESTIMATED CHANGES IN GLOBAL SURFACE TEMPERATURE

Best estimates of increase in global surface temperature in °C relative to the 1850-1900 period average (includes the +1.1 °C since 1850)

Source: IPCC Assessment Report (WG1) p. 88

Scenario	2040	2060	2100
SSP1-1.9	1.5	1.6	1.4
SSP1-2.6	1.5	1.7	1.8
SSP2-4.5	1.5	2.0	2.7
SSP3-7.0	1.5	2.1	3.6
SSP5-8.5	1.6	2.4	4.4

SCENARIO DESCRIPTIONS

Source: IPCC Assessment Report (WG1) p. 88

Scenario (Warning: judgements are mind)	**CO_2 emissions starting in 2015** (rf. p. 88 du Summary for Policymakers, WG1)	**Temperature rise in 2100**
SSP1-1.9 Too optimistic. Net Zero in 2050	Emissions reduction right now	1.4 °C (+0.3)
SSP1.2.6 Unrealistic. Net Zero in 2050	Emissions reduction right now	1.8 °C (+0.7)
SSP2-4.5 A bit more realist	Increase slowly up until 2040; then decrease. Will have decreased 50% in 2100	2.7 °C (+1.6)
SSP3-7.0 Maybe too pessimistic	Will have doubled in 2100	3.6 °C (+1.5)
SSP5-8.5 Non credible	Will have tripled in 2100	4.4 °C (+3.3)

What can be inferred from these tables:

- That in the short term, by 2040, the temperature is likely to increase by 0.4 °C (1.5 °C – 1.1 °C) regardless of the scenario.
- That both SSP1 scenarios meet the Paris Agreement targets with an increase of less than 2° C by 2100. Net Zero by 2050. Very hard to imagine this happening.
- That the 3rd scenario requires a very large amount of clean energy (which emits no CO_2 at all) so that emissions can decrease starting in 2040, that is in 17 years! It should be noted that not all renewable energies are zero-carbon.
- That two of the scenarios where CO_2 emissions double and triple by 2100 are hardly more credible than the two unrealistic scenarios that achieve carbon neutrality in less than 30 years.

IPCC scenario analysis

The good news

Good news, says Pielke![15] Finally, the IPCC recognizes that the SSP5-8.5 scenario is highly unlikely due to recent developments in the energy sector. Unfortunately, though, this statement is deeply buried in the report (it can be found on page 238 of the WG1 report). Here it is:

> "However, **the likelihood** of high-emissions scenarios such as RCP8.5 or SSP5-8.5 **is considered low** in light of recent developments in the energy sector (Hausfather and Peters, 2020a, b)." [16]

While this admission is welcome, critics point out that it should have been made very clear in the *Summary for Policymakers*, the main resource of the media and governments.[17]

Major differences between IPCC and IEA greenhouse gas emissions' projections

A recent study *Distorting the view of our climate future: The misuse and abuse of climate pathways and scenarios*[18] reveals that the GHG emissions projected with SSP5-8.5 have now nothing to do with what was initially projected thirty years ago.

This scenario estimates that the use of coal would increase sixfold by 2100 and that the number of coal-fired power plants in the world would increase from 6000 to 33,000![19] In contrast, projections by the *International Energy Agency* (IEA) show that fossil fuel emissions are about to plateau.[20]

A reference scenario that discredits climate change science

The world's governments, financial, insurance and strategic consulting sectors, to name a few, use the worst-case scenario (RCP8.5 or SSP8.5) to design their products and for regional adaptation planning.[21] The number of mentions in the IPCC report shows that more than half refer to the most extreme scenarios...

Hundreds of research projects and climate studies are published every week using the most improbable RCP8.5 and SSP5-8.5 scenarios as a reference point. According to scientist Judith Curry:

> "The new IEA (*International Energy Agency*) scenarios have rendered obsolete much of the climate impacts literature of the past decade, which have focused on RCP8.5, and the accompanying media reporting. It is now clear that climate impact assessments have been biased in an alarming direction by continued inclusion, and even sole reliance, on

> RCP8.5. The end result is that the scientific literature has become imbalanced in an apocalyptic direction, which is amplified in the media to stoke alarm about climate change."[22]

The candy-rose scenarios

In a good *National Geographic* article, the two SSP1 scenarios are seen as "optimistic projections that reach the Paris Agreement targets: keeping global warming below 2° C."[23] In these scenarios, carbon neutrality is reached in 30 years (in 2050)!

These are nice projections. But to achieve these targets, emissions would have to immediately go down in a very significant way. Otherwise, it is hard to imagine it possible, which means that of the 5 official IPCC scenarios, only one is realistic…maybe.

Overheated climate models and Climate sensitivity to CO_2

If you work for a company and there is a computer system update during the weekend, you surely have noticed that there is always something that goes wrong come Monday morning.

When I learned about all the uncertainties surrounding climate modeling, my first impression was that we were in a "pilot phase." It also came to me that it was downright crazy to believe that computer programs could anticipate the behavior of such a chaotic system. A little like it is ridiculous to believe that we can anticipate the behavior of the stock market. In these types of systems, the slightest change of a variable may have unexpected effects on the whole system.

In the case of climate models, there is an important metric that measures temperature sensitivity to CO_2 emissions. It is called *Equilibrium Climate Sensitivity* (ECS). What happens if you double the concentration of CO_2 in the atmosphere? As already noted, the CO_2 concentration has not yet doubled since 1750 (from 280 ppm to 421 ppm in 2022).[24] But if it doubled … of how many degrees would the temperature increase? How sensitive is the temperature to this doubling of CO_2? This is where ECS comes in, a metric which is at the heart of the climate modeling debate.

If the climate sensitivity (ECS) is high, more global warming can be expected. It helps maintaining a sense of urgency to reduce GHGs. But if climate sensitivity is low, it would mean that scientists who defend this view would be right and aggressive policies to reduce emissions would be more difficult to justify.[25]

Do you think IPCC's science can tell us for sure how sensitive the climate is to the doubling of CO_2? Think again. We are in a sea of uncertainties. While in 2013, the IPCC estimated that the "most likely" value of ECS was between 1.5 °C and 4.5 °C, the latest report now proposes the interval 2.5 °C to 4 °C with a most likely value of 3 °C.[26] Here is a table that shows the evolution of IPCC's statements on *Equilibrium Climate Sensitivity* (ECS).

IPCC REPORT	ECS RANGE	MOST LIKELY VALUE
AR4 2007	2 - 4.5	3 °C
AR5 2013	1.5 - 4.5	None given
AR6 2021	2.5 - 4	3 °C

CLINTEL, *The Frozen Climate Views of the IPCC*
May 2023, p. 95[27]

What happened? The lower bound of the interval went from 1.5 °C to 2.5 °C. And so, the climate would therefore be more sensitive to CO_2 emissions!

An excellent article[28] by Ross McKitrick* helps us understand where the last IPCC estimate (2.5 – 4) came from. To make a long story short, it was based on a study by the Australian Sherwood that concluded that the ECS was between 2.6 °C and 3.9 °C. A few months later, however, appeared a new peer-reviewed study signed by an English mathematician Nicholas Lewis.[29]

After correcting mathematical errors and updating the data of Sherwood's study, Lewis found that the sensitivity range was **1.8 °C et 2.7 °C** with a most likely value of **2.2 °C**. A big difference with IPCC's **2.5 °C à 4 °C** estimated range of values! In Lewis's findings, climate would therefore be *less* sensitive to the doubling of CO_2's concentration in the atmosphere!

	ECS RANGE	MOST LIKELY VALUE
Sherwood et al.	2.6 – 3.9	-
Nicholas Lewis	1.8 – 2.7	2.2
IPCC AR6 2021	2.5 – 4.0	3.0

Overestimating climate sensitivity (ECS) has serious consequences, McKitrick tells us. *The difference between approximations at 2.2°C rather than 3°C is enormous.*

* The same McKitrick that debunked, with Steve McIntyre the *hockey stick* chart.

Several cost/benefit or social cost of carbon studies take ECS into account.

As for the new IPCC SSP models, there is definitely overheating. Dozens of models even produce *more sensitive climates than the upper 4 °C limit*![30] Researchers have tried to find the source of the problem; it seems that maybe the treatment of clouds by climate models would be a factor...

The Sun's role

What role does the sun play in the global warming of the 20th century? This is an issue that was overlooked in the 6th IPCC report, according to the *Clintel Foundation* analysis *The Frozen Climate Views of the IPCC* [31]

If I understand correctly, there are cyclical variations in the sun's rays' intensity. However, according to these analysts, "the IPCC conclusions only take into account the scientific literature that minimizes the role of the sun" and therefore, the relationship between climate and the intensity of solar activity. Some of these studies even make the connection between the warmer and colder periods of the past and solar activity fluctuation. These illustrations are impressive.

Even more recently, a new study suggests that *Global warming could be mostly an urban problem*. Among other things:

> "The study also suggests that the solar activity estimates considered in the most recent reports by the UN's Intergovernmental Panel on Climate Change (IPCC) likely underestimated the role of the Sun in global warming since the 19th century." [32]

While they find that warming and cooling trends in rural data could be explained by changes in solar activity, the study also suggests that warming in urban areas could account for up to 40% of the warming since 1850! That's a lot of warming!

A whole lot is being written on the sun's influence on climate. I personally had the impression that the debate is very much ongoing. For Judith Curry, "the next 20-30 years of observation should reveal a lot about the role of the Sun in climate."[33] Let us hope that financial support will follow since it has become more and more perilous to obtain research grants when studies do not fit the official narrative.

Michael Mann's magic

We've already mentioned the famous *hockey stick* chart in a previous chapter. Let us recall that its author and the famous illustration are what explains the virtual disappearance in the IPCC reports of periods in the past that were just as warm as those of today (especially those of the medieval period).

We've also noted the possible contribution to climate of the Atlantic Multidecadal Oscillation (AMO), a cyclical phenomenon consisting of ocean temperature changes of the North Atlantic Ocean surface.

Well, this time, from computer climate simulations, it is the disappearance of the AMO that Michael Mann announces!

> "We conclude that there is no compelling evidence for internal multidecadal oscillations in the climate system."[34]

In fact, it gets even better. An article published in a scientific journal presents the study with the title *Apparent Atlantic Warming Cycle likely an Artefact of Climate Forcing.*[35] In other words, the AMO is not a factor contributing to climate change but is one of its artefacts!

On her blog, Curry publishes *Canceling the AMO*[36] in which she is impressed. "Wow," she writes. "In one fell swoop, […] pesky problems and debates all go away. Brilliant! Almost as 'brilliant' as the hockey stick."

That made me smile.

CHAPTER 5

No, there is no climate crisis

I found the consensus!

It is in Judith Curry's book that we can find, in my opinion, the best assessment of today's climate science: what we know with a high degree of certainty and what we do not know … yet. If there is a consensus that can be endorsed by a strong majority of scientists, it is this one:

- "Average global surface temperatures have overall increased since about 1860.
- CO_2 has infrared emission spectra, and thus acts to warm the planet.
- Humans have been adding CO_2 to the atmosphere via emissions from burning fossil fuels."

However, these facts, she adds, taken individually or even collectively, cannot answer the following questions:

- "Whether and to what extent CO_2 and other human-caused emissions have dominated over natural climate variability as the cause of the recent warming.
- How much the climate can be expected to change over the twenty-first century.
- Whether warming is dangerous.
- Whether radically reducing CO_2 emissions will improve human well-being in the twenty-first century." [1]

To answer these questions, much more research is needed to detect and assess the natural influences we saw in previous chapters. We are referring here to research related to paleoclimatology, geology, archaeology, or any other science that explores Earth's history.

Also, more and more scientists are openly debating on their disagreements, whether on studies published in scientific journals or about the IPCC reports themselves. This is how I would summarize them:

- The definition of "climate change" in the United Nations Framework Convention needs to be amended. It is misleading and may suggest that all climate change is caused by human activity. However, there are many natural influences that act on the climate, but that science is not yet able to correctly identify, let alone measure.
- The reliability of climate models is questionable, particularly because they fail to accurately reproduce the not-so-distant past; how can they predict what the climate will be in 2100?
- There is overheating in climate models. For some of them, climate sensitivity is even higher than the upper bound IPCC interval. Let's remember that overheating makes it easier to justify and promote a sense of urgency. (Maybe that's the idea?)
- There are significant contradictions between scenario results and empirical data reported in the scientific literature. Moreover, the persistence of the IPCC and policymakers in continuing to use the SSP5-8.5 scenario renders the scientific literature and the media coverage that makes use of it not credible.
- One of the IPCC's missions is to render a comprehensive, unbiased, peer-reviewed literature review. That's not the case. Historically, last-minute studies have been retained and cited in IPCC reports; some have not even needed to be published before being cited. *Cherry-picking* studies and time scales

biases are reported. There is also bias for studies that wholly focus on the negative consequences of warming.

- It is very difficult to justify the danger of climate change based on more frequent and devastating extreme weather events. The IPCC WG1 itself has difficulty detecting such changes. Most of these events have existed in the past at times when CO_2 had little influence.

Finally, if there is a consensus, it is very thin. And with so many uncertainties, I find myself asking "how can one adopt such a pretentiously apocalyptic tone?"

Indeed, when we go back to the 1992 *United Nations Framework Convention on Climate Change*, can we not reasonably doubt the foundation on which the diagnosis was based on (warming is anthropogenic and dangerous) but also the solution (CO_2 must urgently be reduced)? After thirty years of global catastrophic predictions, can we recall that at the time of its adoption, no scientific evidence could support the detection of any climate change, let alone that of anthropogenic change. As for the concept of dangerous warming, why would we prefer the pre-industrial temperatures of the Little Ice Age?

In a system as extraordinarily complex as the global climate, is it really conceivable that CO_2 could act as a control knob that would take precedence over, and neutralize all other natural influences? Let's think about that for a second ... That *one variable*, CO_2, can cause such an apocalyptic disruption?

Claiming that disaster is imminent implies that temperature rises have never occurred in the past at times when CO_2 was irrelevant. It also imposes on us the idea that a warming of a few degrees is necessarily harmful and causes more misfortunes, famines and wars.

To date, history has shown the opposite… The planet is greening and food production is 45% higher than 22 years ago, using less land.[2]

Red-tinted journalism

There is another (and very loud) consensus emerging from these authors' books on global warming. In many respects, it is shared by several other scientists and a growing number of citizens seeking a balanced coverage on climate change. It is the mediocre quality of climate change journalism. Activism and ignorance significantly contaminate what is presented to citizens.

For some of these scientists, this mediocre journalism is the result of an implicit social contract between some scientists, the media, and policymakers. A contract based on major sources of public funding that too often transform journalists into mere spokespersons for political leaders. To grab the attention of readers and listeners, more and more headlines are sensationalized without any kind of nuance. It is as if reporters transcribe "as is" without taking the time to contact other specialists that would offer another angle or provide a counterweight to the official statements of the IPCC, the United Nations, and their governments.

Now supporters, they no longer recognize the biases of scientific journals and prestigious magazines. Here too, they exist and unfortunately have the effect of only lending their publishing platform to the 'acceptable' researchers, the 'acceptable' articles that conform to the 'acceptable' narrative. Fortunately, voices are being raised against the collusion between climate activist scientists and activist journalists. Such an experience is recounted by Pielke[3] on his

Substack when a whistleblower sends him all the emails exchanged between the authors of a study, the editors of an academic newspaper and the very important publishing house.

Nine months after the study was published, *The Guardian* newspaper harshly criticized it, quoting four scientists including the famous Michael Mann. We can guess what happens next. In the end, the paper was retracted. The four scientists who had bypassed the usual peer review process, teaming up with activist journalists, managed to exercise nothing less than censorship.

By the way, the article reviewed extreme weather events in line with the most recent IPCC report (AR5) at the time. One conclusion was as follows.

> "In conclusion on the basis of observational data, the climate crisis that, according to many sources, we are experiencing today, is not evident yet." [4]

The contours of the catastrophist narrative also encourage polarization by creating on one side, a camp that preaches an unverifiable doxa and on the other, a camp that can be described as denialist. Sounds familiar, doesn't it? During the pandemic, editorial writers and columnists identified very clearly for us who the culprits were. Today, they denounce the disease of economic growth[5] and consumerism. The Western way of life is put on trial and the culprits are easily identifiable. In Quebec, for example, it is reported that the wealthiest 1% of the population individually produce nearly 4 times more GHGs per year compared to those who are in the bottom 40%[6].

Climate change is said to be a major cause of health problems and mortality. "Climate change is already killing us"[7] says the Director General of the World Health Organization in a *Foreign Affairs* report. In *Unsettled*, Koonin is outraged.

> "The World Health Organization itself has said that indoor air pollution in poor countries – the result of cooking with wood and animal and crop waste – is the most serious environmental problem in the world, affecting up to three billion people[8] This is not the result of climate change. It's the result of poverty."[9]

Pollution-related deaths are not caused by climate change; it is pollution itself that kills, he adds. Blatant misinformation by WHO management. The example comes from the top…

Closer to home, the excellent observer of the Canadian political scene Rex Murphy points out the blatant activist bias of the CBC.

> "Forty-four years of David Suzuki and not one interview with Bjorn Lomborg or Judith Curry or — as fine a mind on science as the world has to offer — Dr. Richard Lindzen?
>
> All of CBC's environmental reporters — I repeat, all — are environmentalists. Name the one exception. You can't. Unbiased? Mr. Fenlon has a great sense of humour."[10]

As I write this chapter, I see my TV screen splashed with all the variations of red to illustrate heat waves. What used to be seen as beautiful and sweet warm summers turns from a pale orange to red; the heatwaves, meanwhile, turn purple. The message is clear: it's not what it used to be and things are going to get worse!

On July 27, 2023, United Nations Chief Antonio Guterres insists hysterically once again and puts humanity on alert:

> "The era of global warming has ended. The era of global boiling has arrived."[11]

You are warned. The planet is boiling! I don't know about you, but for me, this type of declaration is completely insane.

The consequences of climate alarmism

Fear has always been a formidable weapon. We have seen the damage it caused during the pandemic: in our country, poorly targeted health policies, more vulnerable cohorts neglected, flagrant media unanimism, governments and professional associations complicit in imposing censorship. Now that the planet is boiling, it is hard to imagine the consequences of such irresponsible statements.

Consequences on mental health

> "Katie and Aaron are millennials [....]. They reached adulthood as the world began to realize the catastrophic impact of human beings on the environment. Before getting married in 2017, the couple struggled for almost a decade with the ethical dilemma of bringing another human being to an already overcrowded planet. Katie pleaded to raise a climate ally, someone who would fight for a healthy planet [...]" - *National Geographic*, July 2, 2022."[12]

> "A Montreal district has distributed to its citizens a leaflet suggesting that having one less child would be the best action to fight climate change." - *Journal de Québec*, May 19, 2023.[13]

> "Young Canadians are more eco-anxious than ever. Just under half of the 1,000 young Canadians aged 16 to 25 questioned by researchers at Ontario's Lakehead University believe there is no hope for humanity." - *Le Devoir*, March 20, 2023.[14]

These are not Pakistanis youngsters between the ages of 10 and 19 following the 2010 flood, the worst the country had ever experienced. An event killing 1,700 people, affecting 20% of the territory and destroying 1.1 million homes.[15] In *Le Devoir's* article, these are young Canadians much more influenced by climate doctrine than by their own experience of weather events' ferocities.

This is what apocalyptic rhetoric leads to for many young people: depression, post-traumatic stress disorder, learning disabilities, the feeling of having been betrayed, etc. Very young, they are confronted with images of a crying planet, a screaming planet and a planet destroyed by mankind. Which parent did not notice the contamination of school texts that allude to environmental and catastrophic climate change? The elite's praise of the young Greta Thunberg's adds to the fervor and if in the past, young people wondered why it was important to study physics since they "would never need it", today, the most affected wonder why study at all. At a protest in Denver in 2020, one sign read "We will not die of old age; we will die of climate change."[16]

Public health institutions are also not helping as they should. What are they doing? They are "listening"— The following are "expert" quotes selected by the April 2023 *Canadian Public Health report.*[17]

> "If we don't address capitalism, if we don't address colonialism, racism, the patriarchy, etcetera, we're going to tread water for a long time until we eventually drown [...]"

One of the five guiding messages from the report:

> "The systemic drivers of negative health outcomes and climate change overlap: white supremacy, capitalism, colonialism, and racism must be addressed."

As for what is happening in our crumbling provincial healthcare systems such as its accessibility problems, wait-times problems and staff shortages, we need not worry. *La Presse* reporter Ariane Krol reports that the *Centre intégré de services de santé et de services sociaux (CISSS)* of Laval is tackling its carbon footprint![18] Better still, about twenty doctors suffering from eco-anxiety decided to "get involved, take time and energy to find solutions to self-treat."[19] Let's read that again: "to self-treat" themselves"!

I am being ironic, of course. But this, ladies and gentlemen, is the priority of certain public institutions funded by our heavy contributions to our supposedly universal Canadian and Quebec healthcare system.

Consequences on government's lack of responsibility

The CO_2-centric policy, which relies essentially on the reduction of GHGs to ensure Earth's salvation, offers an invaluable way out for governments. We see it for example, when our water supply systems burst. Or in government's deficient management of banks and flood zones as in the case of the Gouffre River that we saw earlier.

Residences for the elderly do not offer air conditioning in hot weather. Blame climate change. Forest fires, tornadoes, floods: blame climate change. That's what the science says, right? Wrong. We now know that it is not. We understand why the media and political activists encourage this line of thought: for the former, it's easy-breezy journalism and for the latter, governments are cleared of any responsibility if climate change is the culprit of all their mismanagement. But even IPCC's science does not suggest that every extreme weather event is caused by climate change. In fact, it concludes the opposite.

Governments around the world are being exonerated of their responsibilities. Their neglect to maintain aging infrastructure is barely mentioned; better to insist of climate change damages, target capitalism and blame consumerism!

The dominance of climate change as the ultimate culprit of everything that goes wrong therefore exonerates all levels of governance of any lack of wise governance: municipal, national and international. In the worst cases, even famine and scarcity of drinking water are not due to climate change, but to short-sighted land management and irrigation policies, deforestation, outdated urban planning plans, etc.[20] Case in point: British Columbia's Fraser River and the Colorado River in the American Southwest discussed in the chapter on extreme weather events.

Consequences on cost evaluation

In *False Alarm*, Lomborg introduces us to economist William Nordhaus, winner of the 2018 Nobel Prize in Economics. In 1991, Nordhaus wrote one of the first studies on the financial impacts of climate change. Inspired by Nordhaus' work and unlike the CO_2 obsession of the United Nations, the IPCC and the COPs of this world, Lomborg's and the *Copenhagen Consensus Centre's* approach was to evaluate a whole set of policies aimed at improving the human condition.

> "If we truly want to make the world a better place, we have to be very careful that our preoccupation with climate change doesn't distract us from other crucial problems. [...] Fixating on scary stories about climate change leads us to

> make poor decisions. Overspending on bad climate policies doesn't just waste money. It means underspending on *effective* climate policies and underspending on the opportunities we have to improve life for billions of people, now and into the future."[21]

For Lomborg, climate panic has generated something that could be compared to the military complex-industrial: way too much spending on bad climate policies at the expense of public policies that are crucial to the development of human well-being (education, health, pollution, malnutrition, etc.).

A wealthier society will have the means to take an interest in preserving nature, reducing pollution, and replanting its forests, he adds. Otherwise, the obsession with CO_2 mitigation policies, for example, is not only inefficient, but morally reprehensible. Even more reprehensible since the estimated costs of damage related to climate change are very often based on the RCP8.5 (or SSP8.5) disaster scenario which the IPCC itself now describes as highly unlikely.

Besides, over the past 30 years, IPCC emissions' projections have been grossly exaggerated, writes the economist Ross McKitrick in the *Financial Post.*[22] Comparing the 1970s, 1980s and 1990s IPCC emissions' projections with historical data will confirm this. In fact, says McKitrick "if we want to avoid the RCP8.5 future scenario, all we have to do is stop feeding it into climate models, because that's the only place it exists." I smiled at that too.

Despite this, in April 2023, Canadian federal Environment Minister Steven Guilbeault increased the social cost of carbon* from $54 to $247 per tonne! For the taxpayer who

---

* Premises underlying the assessment of the social cost of carbon: 1)

finances governments' climate policies, setting the social cost of carbon is very important. Various regulations, subsidies and energy efficiency measures are justified on the basis of a hypothetical social cost of carbon and on so-called 'savings' achieved by GHG reduction.

However, the methodology used by the Biden administration and Minister Guilbeault to establish the social cost of carbon is seriously flawed. As explained in another of McKitrick's *Financial Post* articles, *Junk Science Week: The Social Cost of Carbon game:*[23]

- The $247 figure referenced by Guilbeault comes from a new report from the Biden administration that tossed out all the previous models, including Nordhaus's.
- These new models persist in using an Equilibrium Climate Sensitivity (ECS) of 3°C, and this, even if the scientific literature shows that ECS is about 2.2 °C or less.
- As for evaluating mortality costs, the study's conclusions also rely on the discredited RCP8.5 scenario, even if RCP4.5 is also used... but somewhere in a 113-page online-only appendix.

After careful study, this is the graph that McKitrick comes up with and that is presented in the *Financial Post* article as well. Notice its title "*The Iffy 'Science' Behind Social Carbon Costs*":

Premises for assessing the social cost of carbon: 1) GHGs contribute to global warming and 2) rising average temperatures increase the number and severity of extreme weather events (which the IPCC has not demonstrated for all events).

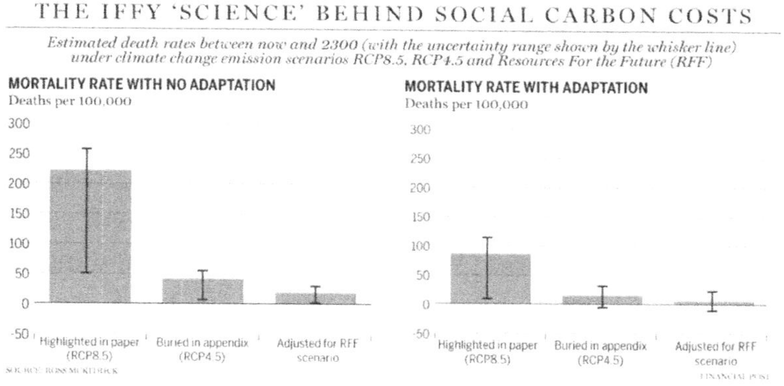

Two things stand out from this chart:

- First, the importance of adaptation-based climate policies. See how mortality rates fall by more than half when there are adaptation policies.
- And secondly, there is even a RCP4.5 scenario that produces near-zero mortality rates if you consider a variant using higher income growth.

Of course, neither Guilbeault's announcement nor media analysts covered anything else but the results based on the apocalyptic RCP8.5 scenario.

Consequences on the environment

There was a time when, in popular parlance, environmental protection was linked to the elements of nature and the diversity of living species. What came to mind when we spoke of the environment were natural landscapes, lakes, and forests. We were concerned with species diversity, air and water quality, flora, and forests, and so on.

Now, by redirecting a significant proportion of investments and subsidies to mitigation and energy transition policies, the "environment" as we used to understand it has taken a back seat.

Ask the man on the street what the biggest environmental problem is, and he will very likely answer that it's climate change. Yet it is air pollution that billions of people suffer from. To the question "What pollutes the world the most?" you can also bet that the answer will be "CO_2, of course!" even though CO_2 has nothing to do with it.

Yet isn't it ironic that nations that miserably fail to manage their own waste believe they can control the climate with a fictitious carbon dioxide control knob?

There are certainly good reasons to invest in clean energy research such as nuclear power, or to switch from highly polluting energy that affects air quality and important sources of drinking water to less polluting energy. However, the temperature fluctuation of a few degrees upwards since the end of the Little Ice Age in pre-industrial times is not one of them.

Many of the problems said to be caused by climate change are actually problems of poor governance or aging infrastructure. As for developing countries, the problem stems from their inability to prevent, warn, and manage the extreme weather events that regularly affect their regions, *with or without climate change*!

In this regard, a passage taken from Judith Curry's book is particularly revealing. Referring to an interview of one of the 5th IPCC Assessment Report's authors, this is Richard Tol with a British journalist:

> "I think that the Fifth Assessment Report in the initial drafts of the *Summary for Policymakers* that actually came through I think very loud is that the new insight [...] is that many of the more dramatic impacts of climate change are really symptoms of mismanagement and poverty and can be controlled if we had better governance and more development."[24] – Richard Tol

Subsequently though, the IPCC modified this conclusion and adopted the traditional, more apocalyptic message justifying the reduction of greenhouse gas emissions. In the end, Tol withdrew his name as a signatory to the IPCC *Summary for Policymakers*...

And yet, life is getting better

There is growing evidence that alarmist fearmongering can be counterproductive. The exaggerations and relentlessness of certain institutions in pushing the UN agenda are less and less well received by the public.

Firstly, no IPCC report forecasts the death of billions of people. The concept of overpopulation strikes the imagination, but the United Nations now projects that "the world's population (now 8 billion) will reach a peak of around 10.4 billion people during the 2080s and will remain at that level until 2100."[25]

Here is some good news as reported by the *Human Progress* organization.[26]

- Between 1950 and 2020, the average inflation-adjusted income per person rose from $4,158 to $16,904, or 307%!
- Between 1960 and 2019, the average life expectancy rose from 50.9 years to 72.9 years, or 43.2%.

- Deaths from inter-state wars fell from a high of 596,000 in 1950 to a low of 49,000 in 2020, or 92%.

- The rates of extreme poverty have plummeted, with the share of people living on less than $1.90 per day declining from 36% in 1990 to 8.7% in 2019. Though the pandemic has temporarily worsened that number somewhat.

- Between 1969 and 2019, the average infant mortality rate per 1,000 live births fell from 89.7 to 20.9 or 77%.

- The literacy rate among men aged 15 and older rose from 74% in 1975 to 90% in 2018. The literacy rate among women aged 15 and older rose from 56% in 1976 to 83% in 2018.

- The chance of a person dying in a natural catastrophe – earthquake, flood, drought, storm, wildfire, landslide or epidemic – fell by almost 99% over the last century.

- Between 1982 and 2016, the global tree canopy cover increased by an area larger than Alaska and Montana combined.

- In 2017, the World Database of Protected Areas reported that 15% of the planet's land surface was covered by protected areas. That's an area almost double the size of the US.

- Good news for the fish: since 2012, more than half of all seafood consumed came from aquaculture, as opposed to the fish caught in the wild.

As for the earth's ability to feed its inhabitants, although *The New York Times* published an article in August 2019 entitled *Climate Change Threatens the World's Food Supply*, Koonin notes that a more careful read of the *IPCC Special Report* would have revealed:

> "[...] that the global production of both crop and animal calories has gone up dramatically since 1960, and that beginning in 1965, more than enough food calories have been produced every year to satisfy humanity's nutritional requirements.

> That ability stems largely from improvements in crop yields. In the fifty years from 1961 to 2011, global yields of wheat, rice, and maize (corn) have each more than doubled, and US corn yields have more than tripled."[27]

Finally, some wild animal species are making a comeback. A *EuroNews* article titled *Wolves, bears and bison: 50 species make 'spectacular' comeback in Europe*[28] reports that animals thought to be on the brink of extinction are making a spectacular reappearance.

The need to get a grip and tone it down

The climate fluctuates. That is a fact. But it has been fluctuating for billions of years. Climate is also unpredictable. Has been for billions of years. Why on earth would this suddenly be seen as a threat to the existence of life on earth? Let's not forget that not so long ago, it was the end of a very cold period, the Little Ice Age. In short, throughout the interglacial period in which we are now, temperatures have fluctuated.

There is no reason to adopt an alarmist or cataclysmic tone. All that it does is signal a strong ignorance of history. It is not constructive. It is moralizing, pointless and irresponsible. We need to pull ourselves together and tone it down.

The voices that argue that there is no "climate crisis" are increasingly heard. The authors featured in this book are of this opinion. Many of these scientists are financially independent; that's no accident; it is an open secret that funding is easier for studies that adhere to the official narrative. Some others are Nobel Prize winners. Many publish books on the subject but are rarely heard of through the mainstream media.

They believe that CO_2 adds to warming, but that this human contribution is at best "modest," insignificant even. As we have seen, that is not the problem. The problem is in the tone.

And this tone is not constructive. In the long term, it will produce indifference, mistrust, and contempt towards climate change *that the alarmists themselves will have created.* Negative feedback to the superlative overkill can already be observed.

Many scientists are disappointed by the narrowness and politicization of the debate, brought about by a consensus that isn't a consensus at all. The result is an oversimplification of both science and solutions. Given the failure of illusory global CO_2 mitigation and energy transition policies, they prefer to develop other, more regional risk assessment and management methods, or opt for adaptation strategies that would enable as many people as possible to adapt to the cyclical moods of the climate.

In September 2022, IPSOS conducted a survey across 5 continents and 30 countries that revealed that "… climate scepticism has been steadily increasing for the past 3 years (37%, +6 points in 3 years)". Further on, it states that the idea that progresses the most is not the negation of a change, but rather the idea that it is "mainly due to natural phenomena as the Earth has always known in its history." There are therefore fewer people than before who believe that climate change is due to human activity (CO_2).[29]

One day, essays will be written to try to understand this moment of climate panic coming from a comfortable West that enjoys a quality of life never before achieved in human history. Indeed, what will be said about the trillions of dollars in research subsidies and costly policies dictated by an unelected level of global governance, that is the United Nations?

In short, blindly, and uncritically promoting the apocalyptic scenario does not qualify as science. In science, skepticism is perfectly healthy. Especially when uncertainties multiply with each new discovery. Science implies that scientists doubt, argue and justify themselves. Let's not spoil their fun by ensnaring them in childish dogmas reflecting our insecurities, or worse, an authoritarian impulse to impose a worldview radically different from the one that has enabled its greatest advances in its history.

After all, mankind has already met some major challenges. There's no reason why our societies shouldn't be able to adapt, if necessary, to fluctuating climates. To imagine that CO_2 is a control knob that will calm the moods of the extraordinarily complex physical elements that surround us is excessively presumptuous.

Especially since, as Bernie Lewin recounts in *Searching for the Catastrophe Signal - The Origins of the IPCC,*[30] this so-called climate science is so very young. Not so long ago, a little over a century, meteorologists believed that a region's climate was stable. Weather events were reported in relation to a stable average that could be relied upon.

At the same time, people exploring the earth's history through its rocks and sediments were discovering enormous changes over tens of thousands and even millions of years. The research of geologists, in particular, led them to challenge the dogma of the day and introduce the idea that climate was not stable, but constantly changing on multiple time scales. From then on, science told us climate indeed changes.

The long-term perspective of these scientists is enlightening and calms the juvenile reflex of believing that the world began on the day we were born. Wiser scientists see climate change as part of the Earth's long history of glacial and interglacial periods and a world of complex influences. In the universe, this ball on which we live has its own dynamic that goes far beyond our little human lives. Ultimately, Planet Earth can be as nourishing as it can be ferociously cruel.

That climate change is now being used to wage war on civilization and human progress is unfairly generating eco-anxiety. As we'll see in another chapter, the CO_2 obsession is a fad. There are good reasons to encourage clean energy research, but the "civilizational transformation" proposed by the UN under the guise of an IPCC science that is only just developing its primer, is not one of them. Generating dogmas, a religion, demons and imposed codes of living may satisfy preaching activists, but it is not a good prescription.

In summary, as we have seen in previous chapters, neither extreme meteorological events nor the danger of a few degrees of warming can justify the alarmism and urgency of adopting political action plans as radical as those proposed by the panoply of United Nations institutions.

Which bring us to another source of influence: political influence.

CHAPTER 6

Science or politics?

The IPCC's political gene

From the first day of its existence, UN political actors have been influencing the IPCC. After all, the "I" in IPCC is for "Intergovernmental". In fact, they are its creators, promoters, and most fervent defenders. Nor has it been long since the human imagination dreamt of controlling the climate. In 1961, for example, US President John F. Kennedy addressed the United Nations General Assembly. "With the cooperation of all nations, it would be possible to predict and eventually control the weather,"[1] he said.

In *Searching for the Catastrophe Signal - The Origins of the IPCC*, author Bernie Lewin recounts some of the events leading up to the creation of the IPCC. Among these, a conference entitled *Study of Critical Environmental Problems (SCEP)* organized by an MIT management professor. On this occasion, more than 70 participants devoted the entire month of July 1970 to the study of critical environmental problems.[2] At the end of the exercise, here is what they concluded about human influences on climate (emphasis added by the author):

> "While it is conceivable that man *may have* had a *small part* in the most recent climate changes we have just described, it is clear that natural causes *must be* sought. In fact, *as has been frequently pointed out*, it will be *difficult to identify* any man-made effect because, first, with our present state of knowledge, *we do not know* how to relate cause and effect in *such a complex system* and, second, man-made effects will

> be *obscured by* the natural changes that we know *must be occurring*."[3]

This quote is particularly interesting. Doesn't it express what many renowned scientists still believe today? At the time, these doubts were considered relevant and rational. Now, they are judged to be the words of heretics and climate deniers.

That said, another event, the World Climate Conference was held in February 1979. Several UN organizations took part: WMO, UNEP, FAO, UNESCO and WHO.* See how the declaration at the end of the session reveals the leadership the United Nations intends to take. Already, we can anticipate what will dominate the discussions: global warming, human influence and carbon dioxide. In the section on 'Climate and the future', Lewin writes, the emphasis was placed on the threat of carbon dioxide emissions:

> "Emissions warming 'appears plausible', it said, because carbon dioxide 'plays a fundamental role' in determining temperature. While this effect is not yet detectable, it should be noticeable 'by the end of the century.'" [4]

The direction was clear: accelerate research into the various aspects of the 'CO_2 problem'. In a bid to move things forward, UNEP Executive Director Mustafa Tolba**

---

* WMO (World Meteorological Organization); UNEP (United Nations Environment Program); FAO (Food and Agricultural Organization of the UN); l'UNESCO (United Nations Educational, Scientific and Cultural Organization); WHO (World Health Organization of the UN)

** The same Mustafa Tolba that had announced in 1982 "that by the turn of the century, an ecological catastrophe would cause devastation as complete and irreversible as a nuclear holocaust."

organized meetings in Villach, Austria, in 1980 and 1985. The result: speculation on global warming and the effects of CO_2 on temperature, based on computer models. The conclusion was that, if current trends continued, the concentration of CO_2 in the atmosphere would double towards the end of the 21st century, and that global warming would be between 1.5°C and 5.5°C.

For Lewin:

> "... the science behind the global atmospheric scares that emerged in the 1970s is remarkable for its detachment from any empirical grounding. [...] In place of any direct validation by evidence, we have, instead, theoretical models. [...] But while they remain theoretical, *they are not science.* Nevertheless, reporters, activists and politicians take the models as supported by the full authority of science, and they then often extrapolate and exaggerate their predictions to conjure up fanciful images of imminent catastrophe."[5]

Never mind that the Austria's meetings report recognized the difficulty in ascertaining the effect of CO_2 on climate because it was difficult to distinguish variations due to natural influences from those due to CO_2. Never mind that the report's conclusions reveal that the alarmism surrounding CO_2 could only be based on vague and various modeling predictions, which were far from perfect. For UNEP Director Mustapha Tolba, there was no doubt. "We have now resolved most of the doubts about the effects of CO_2", he declared. Miraculously, the burden of proof had been reversed, and we could now move on to the stage of public and political action. In Lewin's words, "the planet's thermostat" had been found!

Another interesting passage in Lewin's book relates to the silence of scientists in the face of glaring contradictions between their work and the overblown claims of UN agencies:

> "The silence of scientists on such gross contradictions is an early indication of an unwritten protocol for those who would come to work under the umbrella of the warming scare."[6]

Clearly, no public criticism would be tolerated and any doubts about the role of CO_2 would be hushed up. One single political declaration was enough. The science was settled!

As for scientists' inability to assess how dangerous warming was, the way around the problem was to set a *warming rate* beyond which the environment could not adapt. "Not quite a thermostat, but a throttle", says Lewin. From then on, there was no need to demonstrate the danger of global warming (for the time being). However, as we would later see with the Paris Agreement of 2015, the concept of setting acceptable temperature rise targets would make its mark.

> *So, at the time when the IPCC was created, no empirical evidence of CO_2's role in global warming had yet been found. As for the dangerous nature of global warming, we can understand why the IPCC is so determined to establish a link between an increase in global surface temperature and extreme weather events. They need it badly.*

Moving along, at the end of the '80s, the governments of North America, Germany, Great Britain, Norway and other Western European countries were all enthusiastic about the issue of global warming.

By the end of 1987, the WMO Executive Council and the UNEP Council opted to set up an *intergovernmental mechanism* that would produce climate assessments. This body would report to the Secretary-General of the United Nations and in November 1988, the first session of the IPCC was held in Geneva. Its goal: *detecting* that climate change is occurring, *demonstrating* that the cause is anthropogenic (due to CO_2) and *stabilizing* GHG emissions in order to protect the planet's climate.

Since then…

What was to become of the IPCC is very far from what was envisioned in 1987, says Lewin: a group of fifty climatologists appointed by thirty governments who would produce a document of a few dozen pages intended for all governments through the WMO and UNEP secretariats. "A year later, three very voluminous documents were drafted in the most expensive process ever undertaken by a United Nations agency on a single issue."

Since then, five climate assessment reports have been produced,* the 6th being a summary of the previous five. As environmental engineer Kenneth Green sums up in *Is Climate Catastrophe Really Ten Years Away?*

> "Yet after some 40 years of alarm, despite little or no progress in reducing global greenhouse gas emissions (which have actually increased steadily), we are still the same 10 or so years from climate catastrophe that we were 20 years ago."[7]

* Publication dates: 1990, 1995, 2001, 2007, 2013, and 2021.

In its 35 years of existence, the IPCC has not been without controversy either, including the Climategate mentioned earlier. It also seems that in the last two or three years, renowned scientists have increasingly and openly questioned the IPCC's work and its shortcomings: lack of scientific rigor, politicization of climate assessments, inappropriate alarmist tone, over-reliance on predictive models rather than empirical data, etc.

Many point to the politicization of the IPCC.

The politicization of the IPCC

When we speak of "climate science", we would expect that a group of scientists tasked with assessing it, submit unbiased and exhaustive reports. A transparent and open peer review process is also a hallmark of the scientific process.

The IPCC, which stands out as an authority on the subject, is nothing of the sort, according to Jason S. Johnston in a paper entitled *The Hand of Government in the Intergovernmental Panel on Climate Change*:[8]

> "The IPCC is not and has never been an objective science assessment organization. It was created by and has always been controlled by the governments of countries that perceive political benefits from international regulatory action to reduce greenhouse gas emissions." [9]

In fact, even if thousands of researchers contribute to climate science, it's true that the IPCC itself is not a scientific organization; it's an administrative organization created by two institutions under the umbrella of the United Nations: the World Meteorological Organization and the United Nations

Environment Programme*. Participants are delegates from the 195 Member States. The "IPCC Bureau", which supervises the work of the working groups (e.g., WG1, WG2, WG3), is made up of 34 members, most of whom are also elected by the IPCC's governing bodies.[10]

"Since its inception, the IPCC's mission of 'assessing' climate change has been inexorably linked to the goal of future international policy action on climate," says Johnston. By entrusting politicians (and not scientists) with control over all aspects of report production, the IPCC gives political direction to its research work.

Other scientists are equally dismayed by what the IPCC has become. Here's what I gather from their impressions.

- The politicization of climate science became institutionalized the day the United Nations Framework Convention imposed a political agenda on it. From then on, research into natural climate variability and the potentially beneficial effects of global warming were marginalized.[11]
- In reality, the IPCC's mission is not to offer an exhaustive assessment of climate research; it is to identify and encourage research that supports the anthropogenic global warming narrative, and consequently its funding.
- Its work does not aim to demonstrate that warming is anthropogenic, nor that it is dangerous. The United Nations Framework Convention on Climate Change says so! Even if, at the time and even today, human influence is "modest," and even if the danger of a few degrees of warming is far from being proven.

---

* Better known as WMO – World Meteorological Organization et UNEP – United Nations Environment Program.

- Nor is its operation based on a rigorous, transparent scientific approach challenged by independent reviewers. Although the IPCC claims to be politically neutral, its aim is to defend and promote the political preferences and agendas of its leaders, even if it means intimidating those who point out shortcuts and biases.
- Finally, the IPCC's final products are not synthesis reports written by the researchers and scientists themselves. They are the result of a line-by-line revision of a first draft that may, in the end, undergo significant alterations by the organization's highest political bodies.

A disturbing peer-review process

Confidence in a review process by external, independent experts is essential to ensure the credibility of research. However, the process of reviewing IPCC reports is completely different from the usual processes of scientific journals. In this case, reviewers only must ensure that the authors *take the comments into consideration, but nothing more*. Authors can reject them without explanation or possibility to appeal.

To be truly independent, reviewers must be chosen by an individual or organization that was not involved in drafting the report. Instead, they are chosen by the co-chairmen who are responsible for the reports.[12]

In Koonin's *Unsettled*'s chapter *Saving the Scientific Method,* he highlights the importance of validating results.

> "Independent reproduction or refutation of the results by other researchers conducting their own studies is a much stronger guarantee of correctness." [13]

There are many ways of corrupting the objectivity of the process and the product. In order to receive the seal of approval of all IPCC member states, the *Summaries for Policymakers* are heavily influenced, he notes, sometimes even drafted by governments themselves whose interest lies in promoting one policy rather than another.

One of these controversial episodes involved the *IPCC's 2nd Assessment Climate Report*, published in 1995, says Lewin. IPCC leaders and some of their supporters were even invited to US Congress sessions to explain themselves. The reason had to do with an inconsistency between the final version of a chapter and the version that had been published. As it happens, this chapter dealt with the *detection of climate change and its attribution to human activity*.

This inconsistency was noted by the highly respected scientist Frédérick Seitz and was the subject of the *Wall Street Journal* editorial of June 12, 1996, entitled *Major deception on global warming*.

> "This IPCC report, like all others, is held in such high regard largely because it has been peer-reviewed. That is, it has been read, discussed, modified, and approved by an international body of experts. These scientists have laid their reputations on the line. But this report is not what it appears to be – it is not the version that was approved by the contributing scientists listed on the title page. In my more than 60 years as a member of the American scientific community, including service as president of both the National Academy of Sciences and the American Physical Society, I have never witnessed a more disturbing corruption of the peer-review process than the events that led to this IPCC report."[14]

A few days later, writes Lewin, the scientific journal *Nature* reported on the study which, incidentally, *had not even been published* at the time of the IPCC report. Indeed, it was published in July 1996, *several months after serving as the centerpiece of the 1995 IPCC Summary*. As expected, its final version claims to have detected that warming is *partly due* to human activities, but it points out that *many uncertainties remain*, particularly those related to natural variability. In fact, the title of the study does not advertise a scientific breakthrough, but rather "research on human influence." In addition, *Science* journal's science reporter quotes two other scientists: one said he wasn't entirely convinced that the GHG signal had been detected; the other had demonstrated that warming could be attributed to natural rather than human influences.

Unfortunately, this kind of discrepancy between the scientists' work and the politicians' review is not uncommon. In fact, I had to pick and choose among those I found most disturbing, as some researchers no longer hesitate to openly report the phenomenon. In *A Tip from an IPCC Insider*[15], for example, Pielke recounts on his Substack how a statement on the detection of category 3 to 5 cyclones can change from "a trend *is not detected* in the last 40 years" to "*it is likely* that the overall proportion of category 3 to 5 cyclone instances and the frequency and intensity of events have increased overall in the last 40 years." Note that the word "likely" is important; it indicates that the probabilities of this statement being true are between 66% and 100%. Furthermore, after verification, Pielke confirms that the final (and false) information was inserted into the reports *outside* the normal revision process.

Let's recap:

- Some research is excluded or ignored if it calls into question the narrative of anthropogenic warming, while others (maybe even not yet published) may be added outside the normal review process.

- The authors responsible for the reports produced by the working groups are appointed by members of the IPCC Bureau, themselves appointed by higher political authorities.
- Summary reports are validated line by line, not by scientists but by political representatives of the member states.
- Certain statements in the summaries in no way reflect the content of the report they summarize.

An increasingly clear drift towards politicizing the IPCC

The alarmist fanfare of UN representatives has been particularly loud since the release of the 6th IPCC Report in 2021. Examples include UN Secretary-General Antonio Guterres's hovering declarations: "time bomb," "highway to climate hell," "climate catastrophe," "collective suicide," "to waste time is to perish," and the latest from August 2023, "the era of global boiling is upon us."

And yet, in the latest 2021 report, the most extreme scenarios were deemed highly unlikely. The most recent Conference of the Parties (COP27) was even working with a scenario that projects +2.5 °C* by 2100, instead of the 4 °C to 5 °C of the most pessimistic scenario! Doesn't this give us reason to play down the UN Secretary-General's statements?

---

* In reality, +1.4 °C since temperature has already increased by 1.1 °C since the pre-industrial era.

Unfortunately, not. The IPCC persists in its climatism, using its status as a "scientific authority" to promote a prefabricated political consensus.[16]

Again, economist and legal scholar Jason S. Johnston's opinion of the IPCC, in *The Hand of Government in the Intergovernmental Panel on Climate Change*,

> "[The IPCC] is and always has been, a political not a scientific organization. It is not an impartial observer of scientific progress: it is an advocate for one side in a longstanding policy movement [...] that has diminished the panel's ongoing relevance. Ironically, however, the IPCC approach strengthens the need for an assessment process that would actually do what people have long, and wrongly, assumed the IPCC does, namely, provide a balanced and accurate assessment of the scientific literature on climate change."[17]

The political agenda of the United Nations and of the IPCC

But what *is* this political agenda shared by the IPCC's core leadership and set out in the *Executive Summary* of the latest 2021 report? Two key words: "transformational change." And what does this "transformational change" refer to? Roger Pielke Jr. has gone through the documentation of the IPCC Working Groups 2 and 3 reports to find out. Here are a few excerpts from these reports, which he presents in a text entitled *The Political Agenda of the IPCC*[18] published on May 15, 2023.

- While transitions involve 'processes that shift development pathways and reorient energy, transport, urban and other subsystems, transformation is the resulting fundamental reorganisation of large-scale socio-economic systems'.

- An inner transition within an individual typically involves a person gaining a deepening sense of peace and a willingness to help others, as well as protecting the climate and the planet.... Examples have also been seen in relation to a similar set of inner transitions to individuals, organisations, and societies, which involve embracing post-development, degrowth, or non-material values that challenge carbon-intensive lifestyles and development models...

But why are these views being used to frame a scientific assessment, asks Pielke?

Also, this agenda of "transformational change" is a political drift from how the term was used in the previous 2014 Report, he says. At that time, "transformation" was limited to technological options for emissions reductions, not a complete change of all our societies' social and economic systems.

This vision, shared by influential members of the IPCC, reveals the instrumentalization of climate science to promote degrowth and foster the rejection of the developed world's way of life. Climate change becomes, in fact, a lever for transforming society. The alarmist and apocalyptic tone of a cataclysmic future thus conditions a sense of urgency. Every ten years, "we must act now," despite data showing that the IPCC has vastly overestimated global warming and the danger of extreme weather events. According to Pielke, the politically symbolic jargon of "transformational change" has indeed become the dominant frame of reference for the IPCC's 6th Report particularly insensitive to the reality of billions of people around the world who have quite different priorities. Ultimately, he believes, the IPCC must choose: a scientific organization or one that uses science to achieve political goals? He adds:

"The political agenda of the IPCC reads as if it was developed by wealthy American and Europeans academics. The billions of people around the world who may lack energy services or enough food probably would welcome an agenda of change.

Instead, the IPCC emphasizes transformational changes in the lifestyles of ordinary people in rich countries, for instance, the recent Synthesis Report explained: 'Many mitigation actions would have benefits for health through lower air pollution, active (e.g., walking, cycling), and shifts to sustainable healthy diets.'"[19]

Science + politics = Politics

For some, the IPCC's main *raison d'être* remains to assess climate science on a periodic basis. To ensure the integrity of the research, they call for a review process that is open, transparent, and independent of political agendas.

For others, the IPCC's mission is enshrined in the *United Nations Framework Convention on Climate Change*. It is both scientific *and* political. Ultimately, the research is aimed at convincing people that global warming is man-made and getting the world's governments to adopt policies to reduce greenhouse gas emissions.

Finally, for others still, the IPCC does not go far enough in integrating social sciences. For Lidskog and Sundqvist, for example, the fight against climate change must now involve the "transformational change" of societies. In an article published in September 2022, they deplore the fact that the concept of "transformational change" is not sufficiently understood. In their view, this approach has become necessary because the actions taken to date have been insufficient, inefficient and too limited. So, even if the latest

IPCC report of 2021 introduces the concept, in their view, it's not enough:

> "If the IPCC wants to be policy-relevant, it must adopt a wider and more nuanced understanding of transformative change when developing pathways and conceptualize society as more than just a set of conditions that enable or restrict technological innovation."[20]

Make of that what you will... but for me, I understand the request, and it's a simple one: that the social sciences be given an importance equivalent to that given to the natural sciences in the IPCC assessments, and with the necessary research budgets, it goes without saying...

Thoughts...

In short, it will be interesting to see what happens to the IPCC in the years to come, and whether it can regain its scientific relevance. For the time being, its catastrophist tone is counter-productive, and its "science" suffers from the absence of empirical data in favor of barely credible modeling. What's more, the botched peer review process reflects the dogmatic nature of the IPCC, which tolerates neither external criticism nor deviation from groupthink.

After 35 years of reports, alarmism and controversy, it would be interesting if the United Nations and the IPCC changed their tune. On July 26, a new IPCC director was appointed to lead the 7th climate assessment cycle. His name is Jim Skea,[21] a researcher in the fields of energy, climate change and technological innovation. He was co-chair of IPCC Working Group 3, which focuses on greenhouse gas reduction policies. He was also on the scientific board of the *Special Report (SR15) on global warming of 1.5 °C.*

In the press release announcing his appointment as head of the IPCC, here's how he sees his mandate.

> "Climate change is an existential threat to our planet. [...] In this, I will pursue three priorities – improving inclusiveness and diversity, shielding scientific integrity and policy relevance of IPCC assessment reports, and making the effective use of the best available science on climate change. My actions as the Chair of the IPCC will ensure that these ambitions are realized."[22]

Skea is therefore a man who knows the IPCC 'machine' well; he's from the 'it's now or never' school of thought. Some researchers found the former chairman "bland and non-existent", reports *FranceInfo*[23]. Imagine... And then, in an interview with Michael Liebreich[24] two years ago, Skea shows that he very well knows who his customers are. "Remember, he tells Liebreich, the 'I' in IPCC stands for 'intergovernmental'."

As for the future of the IPCC, let's look ahead. The first path is that of an IPCC that resists all criticism. We change nothing; we gradually sink into rhetoric and transformational politics.

The second pathway is where the IPCC makes amends and returns to what scientists believe it should be doing: assessing the state of the climate and issuing a set of scenarios based on GHG emissions and certain social parameters (e.g., demographics). The adoption of rigorous review processes by independent, external scientists is essential if we are to regain a minimum of scientific credibility.

In my opinion, however, the most likely scenario is that of a gradual retreat by member states. As the IPCC acts and communicates politically, it runs the risk of encountering obstacles to its drive for civilizational transformation. It is also possible that its recommendations will be ignored. It's easy to imagine, for example, that countries constituting major economic forces might balk at the excessive costs of the recommended policies. Openly or in silence, they will withdraw from mitigation and energy transition agreements. In fact, this is already going on, as in the case of China.

Think of the new geostrategic considerations of energy non-dependence expressed by European powers following the Russia-Ukraine war. Or the Brexit, as an example of disengagement with supranational bodies.

Already, we hear that some countries are permitting new hydrocarbon developments and pipelines and expressing renewed interest in nuclear power. Some others, having relied heavily on the combination of solar and wind power, are now realizing the limit these may represent in the overall energy portfolio, i.e., at most 15% to 20%. The populations of countries such as Germany and France have suffered from what I would call the 'Great Experiment in solar and wind power'. Under the instructions of unelected UN organizations and large consulting firms, some national governments have shut down functioning nuclear power plants. It will be interesting to see how the energy issue plays out in the next election cycle. Will people continue to vote for solar and wind-based energy?

As for populations, especially in the West, they have realized that their own GHG reduction targets are insignificant compared to what countries like China and India produce.

Resistance could also take the form of the rejection of a transformative socialist program associated with the left. The time is not far off when the IPCC will be associated with a leftist *wokist* program. Expect right-wing and center-right parties to campaign directly against unelected bodies seeking to interfere in the local governance of their countries.

Although the population says it supports the intentions to develop green energies, there is a growing opposition to the imposition of new environmental taxes, to electric vehicle subsidy policies paid for out of the taxes of people who can't afford to buy them, to the development of public transport at astronomical cost, and to the imposition of constraints on people who wish to travel (number of airplane trips, cruise ships banned from docking in certain countries, etc.) Confidence is eroding. Concepts such as the ecological footprint turn off a middle-class population that anticipates a series of bans that will limit its choices (e.g., natural gas stoves, etc.).

Soon, if they haven't already, the elites will see even more of the anti-elitism of their own making

By blindly adhering to and prioritizing policies that are too costly, at the expense of priorities closer to the people (health, education, infrastructure), these media-political elites will be further discredited. From the moment we touch on everyday life and curtail citizens' freedoms, it is elected governments that pay the price, not global organizations such as the United Nations, the IPCC or the WEF (World Economic Forum).

The population remembers. The management of COVID hurt. With the snap of a finger, we realized that politicians could interfere in people's lives under abusive pretexts. In a few years, an assessment will be made. Already, reports confirm that popular intuition was in many ways wiser than that of the politicians who capitalized on fear and imposed authoritarian and incoherent measures. Perhaps the COVID episode will be remembered in the future as the one that made a still too naïve section of the population realize that a healthy mistrust is called for when it comes to judging the "good intentions" of governments, especially when they claim to ensure "health security" or avert an "existential threat" with policies of "carbon sanitarism."

Finally, how will we talk about this episode of climate panic in 10 or 20 years' time? The politicians and researchers associated with the IPCC may well be embarking on a 7th cycle of climate assessment, but perhaps the policies of degrowth and wokism associated with it will have the better of them by the time the next report is tabled...

CHAPTER 7

CO_2 and the Net-Zero craze

This chapter will attempt to answer some of the most frequently asked questions about climate change. Is CO_2 itself dangerous? How much CO_2 is in the atmosphere? What are the links between fossil fuel emissions, CO_2 and global warming? Is there really a danger of a climate crisis? Is achieving Net-Zero, the much-vaunted carbon neutrality even realistic?

Contrary to the claims of IPCC leaders and the mainstream media, consensus is a long way off. In fact, I found that three groups of scientists propose different visions: the *CO_2-alarmists*, the *CO_2-lovers* and the *CO_2-realists*. On each issue, opinions often diverge, rarely converge. A table summarizing the positions of each group is presented at the end of the chapter.

In this chapter, I'll also look at the origins of the fossil fuel mitigation (reduction) project. How did the 2 °C and 1.5 °C targets come about? What role does economic cost/benefit analysis play in the IPCC's policy recommendations? A brief review of current events will illustrate the Net-Zero utopia, while the project is confronted with geostrategic issues and the public's response.

Finally, the chapter concludes with our authors' views on the feasibility of carbon neutrality.

Linking GHG emissions and CO_2

The culture of climate change catastrophism of the last 50 years has not given greenhouse gases (GHGs) and CO_2 a good name. Although they occupy a tiny fraction of the atmosphere, they grab all the attention. Yet were it not for the insulating layer provided by GHGs, the earth's average surface temperature would be -18°C instead of +15°C. Here are some figures on atmospheric composition from Koonin's Unsettled.[1]

ATMOSPHERIC COMPOSITION

99%	Nitrogen (78%) + Oxygen (21%)	
1%	Argon	**Largest part of the remaining 1%**
	Water vapor	**0.4% of the atmosphere**; 90% of the atmosphere's ability to intercept heat
	CO_2	**0,0417% of the atmosphere**; 7% of the atmosphere's ability to intercept heat

You read that right. CO_2 currently accounts for only 0.0417% of the atmosphere, but it is a GHG and represents 7% of the atmosphere's capacity to intercept heat. Since 1750, its concentration has risen from 280 ppm (parts per million) to 417 ppm, and in recent decades it has continued to increase every year.

For the vast majority of scientists, there is no doubt that this increase in atmospheric CO_2 concentration is largely attributable to fossil fuel emissions. There's a real consensus here among *CO_2-alarmists*, *CO_2-lovers,* and *CO_2-realists*. In fact, even CO_2-lovers such as Patrick Moore of the *CO_2 Coalition* acknowledge that mankind has participated in the 50% increase in the concentration of CO_2 in the atmosphere (from 280 ppm in 1750, to 310 ppm in 1960, to 417 ppm in 2022).

But unlike the *alarmists*, for him "it's one of the best things that has happened in the history of life and humans."[2] Why? Because CO_2 concentration drops dramatically during glacial periods, and the danger is that it could fall below 150 ppm during the next one, a level below which terrestrial plants could not survive. "No other species could have done this: find fossil fuels!" believes Moore. On the other hand, he admits, we are currently producing more than we need.

While all three groups agree that emissions are contributing to the increase in the concentration of CO_2 in the atmosphere, their *perception of the danger* of this increase, ranges from 'imminent danger of catastrophe' (*alarmists*) to 'very good news' (*lovers*).

For *realists*, the danger lies in reaching a concentration too high for humans. That said, a level of 1000 ppm would not be reached for another 250 years, says Koonin. Until then, the three groups hope that reliable, safe, clean, and affordable energies will be developed in the coming decades.

Linking GHG, CO_2 and global warming

When it comes to the relationship between CO_2 and global warming, things get a little trickier.

For the *alarmists*, the solution is simple and begins with the desire to eliminate fossil fuels altogether. Their real goal: to set in motion an enormous energy transition and socio-economic transformation. For them, since CO_2 acts as a greenhouse gas, they are absolutely certain that CO_2 is the *main cause* of global warming. What's more, CO_2

serves as a lever for the civilizational transformation to which they subscribe.

CO_2-Alarmists:
Fossil fuels → CO_2 → Global warming

For the *lovers*, the increase in emissions has little, if anything, to do with the increase in global warming. According to them, there have always been temperature fluctuations. What's more, according to some of them, the climate began warming more than 150 years *before the use* of fossil fuels. As for the rate of warming, "it has not changed in the last 150 years, while emissions have increased at an exponential rate, they say."[3]

Former Greenpeace founder Patrick Moore is adamant. He "firmly believes that the future will show that this whole hysteria over climate change was a complete fabrication."[4] "No credible scientist would claim that the correlation between rising CO_2 and rising temperature over a tiny time interval (since 1850) is proof of a causal relationship. A correlation is not proof of causation."[5] In short, there would be no evidence in the world of actual observations that CO_2 emissions are the dominant factor in the very slight warming observed since the end of the Little Ice Age (1850).

What's more, scientists William Happer of Princeton University and Richard Lindzen of the Massachusetts Institute of Technology (MIT) argue that fears of a runaway rate of global temperature rise are unjustified. In a paper entitled *Challenging Net-Zero with Science,*[6] the authors explain that the higher the concentration of CO_2, the smaller the warming effect of additional CO_2. At current CO_2 levels, they say, "emissions from fossil fuels will have little impact on global warming. There is therefore no climate emergency or threat," they conclude. It's as if CO_2 has reached a saturation point at which the warming effect is lessened.

I admit that this seems contradictory. Does CO_2 have an effect on global warming or not? To this, their answer is, "Yes, but there comes a time when adding CO_2 has less effect."

> CO_2-Lovers:
>
> *Fossil fuels → CO_2 → For some, little if anything to do with global warming.*
>
> *For others, the warming effect of CO_2 has reached a saturation point in the atmosphere. At this point, adding more will have less and less effect.*

Finally, for the *CO_2-Realists*, "human-emitted CO_2 is a relatively small add-on to a vast natural cycle of carbon moving among the earth's crust, oceans, plants, and atmosphere [...] and despite the precision claimed by climate models, the impact of this on the climate is highly uncertain."[7] The question is therefore not a simple one, and the answers are still being studied. So, to find out the specific effect of added CO_2 on climate, the problem is that there are so many other influences (climate forcings), both natural and human, that contribute almost as much to global warming as CO_2 from human activities, argues Koonin. For the realists, there are many uncertainties and disagreements, not only linked to the complexity of the systems, but also due to the immaturity of climate models.

It may even be possible that the presumed contribution of CO_2 to global warming *is less than the margin of error* for measurement of the effect of water vapour! "Is that true?"

Jordan Peterson asks scientist Richard Lindzen in a podcast[8] "Yes, it is true." So "we are planning on spending two trillion dollars[9] to remediate a problem whose magnitude is so small that it can easily be hidden within another measurement error on the water vapor front?" he asks again. "I think so." replies Lindzen very calmly. "Boy, that's really sad," says Peterson. Very sad indeed.

CO_2-Realists:

Fossil fuels → CO_2 → Modest contribution to global warming. Still too many uncertainties about external and internal climate influences to make a more precise assessment. Acknowledge saturation effect.

Speaking of uncertainties, another good example of where science stands is described in an article[10] that reports on tools for measuring the troposphere's temperature. This layer, located 1 to 10 kilometers above the earth, is a prime location for assessing the effects of greenhouse gases, since it is not influenced by the effects of urbanization. It's a well-known fact that it's hotter in the city than in the country. This article discusses satellites, weather balloons and comparisons between observed data and climate models. Several independent teams review, correct and detect biases, but in the end, the most recent methods and data confirm that the warming of observed temperatures is half that predicted by climate models.

A very interesting conclusion indeed, but what we can also understand from reading the article is that scientists are still fine-tuning their methods of observation in the real world. So, once again, it really could be too early to conclude that a cataclysm will occur based on a science that is still evolving.

The political Net-Zero project

Still with the aim to "strengthen the global response to the threat of climate change," Article 2 of the Paris Agreement[11] seeks to limit the increase in global average temperature to 2 °C and even 1.5 °C compared to pre-industrial levels.

For the IPCC, CO_2 therefore transcends all other influences on climate, an approach perfectly consistent with the *UN Framework Convention on Climate Change*, the canonization of political 'consensus' and the ensuing scope of IPCC's work. For the United Nations and the IPCC, man can control the climate, and CO_2 is a lever that enables him to do so.

Since the Paris Agreement, this CO_2 reduction goal has been transformed into a huge international project involving a *Net-Zero Coalition*[12] of governments, cities, businesses, financial institutions, and others. We are well and truly in the realm of United Nations governance, although responsibility for meeting the targets lies with national governments.

In more practical terms, to stay below 1.5°C, for example, the world will need to cut GHG emissions by 45% overall, compared with 2010 levels, by 2030! To achieve Net-Zero by 2050, the International Energy Agency (IEA)* produced a roadmap in May 2021 to guide governments in their decisions.[13] One of the essential conditions for the project's success is that *all governments participate.*

---

* The International Energy Agency was created in 1974 in the wake of the 1973 oil crisis.

Here are some of their guidelines:

- From 2021 onwards, *no new fossil fuel development projects* (oil, natural gas, coal) may be launched.
- By 2030, 650 gigawatts of solar energy will have to be added annually, which is equivalent to adding *each day* the largest solar farm in existence today.
- By 2030, sales of electric cars *should rise from 5% to 60%.* By 2035, 50% of heavy-duty vehicles will have to be electric.
- By 2050, a *complete transformation* of the way we produce, transport, and consume energy will have to be undertaken.
- In 2050, about 90% of electricity will have to come from renewable sources: 70% *from wind and solar power,* the rest from nuclear power plants.

A glance at the International Energy Agency's roadmap quickly reveals just how unrealistic it clearly is. It's surprising that it isn't more contested, that it isn't the subject of more feasibility analyses that would consider geostrategic imperatives and the lessons learned from experimenting and leaning on solar and wind power.

And then, in the end, we ask ourselves what exactly was the point of making such profound and radical changes? Where did these 2 °C and 1.5 °C targets come from anyway?

Science-based or political targets?

To find out, it's worth going back in history to understand how we came to set such targets, which after all, call for a radical transformation of our societies. Doing so, we would discover that these targets *are not the result* of rigorous studies that objectively demonstrate the advantages and disadvantages of a temperature rise beyond these targets.

The first version of the 2 °C story dates back to the 1970s and the temperature projections made by economist, Yale University professor and Nobel Prize winner William Nordhaus. In a study published in 1977 entitled *Strategies For The Control of Carbon Dioxide,*[14] he estimated that an increase of 2 °C to 3 °C would take temperatures outside the range of fluctuations observed over hundreds of thousands of years.[15] As for CO_2, a first approximation leads him to suggest that a reasonable target would be to double its concentration (from 333 ppm at the time to 666 ppm!). In his view, *CO_2 is not toxic to man* as long as its concentration does not reach 20,000 ppm! Further on, however, he acknowledges that the process used to establish such targets is 'deeply unsatisfactory' both empirically and theoretically.

A second version of the 2 °C story was probably the most crucial to the future of the UN project. An article entitled *Three Views of Two Degrees*[16] published in 2010 recounts the influence of climate physicist Hans Joachim Schellnhuber of the *German Advisory Council for Global Change (WBGU)*, on the then German Environment Minister Angela Merkel.

In 1995, the WBGU adopted Nordhaus's concept of historical temperature intervals, but added a dramatic tone to it. "If the lower or upper limits of the interval are exceeded, *dramatic changes* in ecosystems could occur." That year, Merkel chaired the first session of the IPCC Conference of the Parties (COP) in Bonn. And then the influence spread. In 1996, the Council of the European Union officially adopted the 2 °C target. Suggesting that there was a limit beyond which significant risks would rapidly arise had encouraged the talk of "climate catastrophism." A few years later, in 2017, Schellnhuber received the *Blue Planet*[17] award and was recognized as the 'father of the 2 °C limit.' That year, Angela Merkel was Germany's Prime Minister.

Finally, a third version proposes that these targets are essentially the result of the politicization of climate science by member states with unrealistic demands. Suffice it to recall that in 2009, the UN target was a CO_2 concentration of 450 ppm, which corresponded to a temperature rise of between 2 and 2.4 °C by 2100.[18] In 2015, with the Paris Agreement, the target became 2 °C. In 2018, the IPCC governments commissioned a *Special Report on global warming of 1.5 °C,*[19] and today the new reference target is 1.5 °C.

Would this 1.5 °C target prevent major economic and social losses? Was it based on an exhaustive review of the economic literature on the subject? Or on a formal cost-benefit analysis? Was it in line with what the IPCC had stated in the previous assessment report of 2014? A July 2021 study by economists Murphy & McKitrick answers these questions. Entitled *Off Target - The Economics Literature Does Not Support the 1.5* °C *Climate Ceiling,*[20] it shows that none of this was the basis for the work of the IPCC team responsible for the *1.5 °C Special Report.*

Firstly, this *Special Report* states that, given the difficulty of assessing the impacts of climate change, "cost/benefit analysis are difficult to justify and therefore have not been used in this report." In short, no assessment of the costs/benefits of a further reduction (from 2 °C to 1.5 °C) is offered.

What about the economic literature that inspired the IPCC? It turns out that their *Special Report* was made public on the same day that economist Nordhaus became co-recipient of the Nobel Prize for his economic work on climate change. Curiously though, report Murphy & McKitrick, the new Nobel Prize winner's 2016 studies did not inspire the IPCC's 2018 studies. Ironically, Nordhaus concluded that the consequences of imposing a 1.5 °C limit would have been so severe that humanity was better off doing nothing at all to contain global warming (you may read that again). In fact, Nordhaus argues that the optimal target for climate policies is 3.5 °C, 2 degrees well above the mythical 1.5 °C.

Was the *Special Report* in line with what the IPCC had stated in the previous 5th Assessment Report in 2014? Far from it. Whereas in 2014, their advice reflected the economic literature of the time, the new team assembled for the 2018 *Special Report* discarded this consensus (and indeed, the Nordhaus report) and selected from the available literature three brand-new studies published that same year. So, from the 2014 consensus…

> "For most economic sectors, the impact of climate change will be small relative to the impacts of other drivers (medium evidence, high agreement). Changes in population, age, income, technology, relative prices, lifestyle, regulation, governance, and many other aspects of socioeconomic development will have an impact on the supply and demand of economic goods and services that is large relative to the impact of climate change." – Executive Summary, Chapter 10, WGII.

… we go to a new target based on "a small number of marginal studies which, using dubious methods criticized by other researchers, produce very high estimates of damage.[21]

On this, Murphy & McKitrick conclude that :

> "Canadian policymakers and the public should be wary of pursuing aggressive climate targets, in particular the 1.5°C ceiling, when such goals have been derived politically, not scientifically."[22]

So, what happened between 2009 and 2018? Some believe that there has been a "political calibration" of science and politics.[23] After all, climate modelers respond to their clients' requests, and it just so happens that these clients are the governments of the IPCC member states. However, some modelers just don't buy it.[24]

Now new fields of research are emerging: the legitimizing, through "science," of the IPCC's unrealistic targets, and the study of the dilemma facing climate modelers: science or politics?[25]

Realistic targets?

Now that we have an idea where these targets came from, are they realistic or just wishful thinking? In an interview with Peterson, Koonin doesn't buy it. "I think we're at a stage now where we're realizing how difficult it's going to be, I'd say almost impossible, to reduce emissions. It looks like Net-Zero by the end of the century just isn't going to happen."

How *can* we realistically expect to eliminate CO_2 emissions by 2050 to meet the 1.5 °C target of the Paris Agreement? Energy demand will increase by 50% over the next thirty years, partly due to demographics, but also to growing demand from developing countries. In 2021, Standard & Poor's (S&P) predicted that all fossil fuels would experience rising demand by 2050.[26] And that was *before* the war in Ukraine! Look at what's happening in the world today:

- *L'EXPRESS* reported on August 1, 2023, that an *International Energy Agency*'s report noted that "global oil demand is expected to reach its 'highest level ever' in 2023."[27] A few weeks earlier, the same newspaper published *Pétrole : pourquoi Shell ne réduira plus sa production d'ici 2030* (Oil: why Shell won't be cutting production by 2030).[28]

- According to the 2018 *Special Report on global warming of 1.5 °C*, the IPCC itself warns that "at the current rate of emissions, the 1.5°C threshold will be crossed between 2030 and 2052 and warming will exceed 3-4°C by 2100."

- In November 2022, *The Economist* published a special report on the anticipated failure to reach the 1.5 °C target. The cover story read *Say Goodbye to 1.5 °C. Now is the time to be more realistic*, said the magazine's editorial team.[29]

- In October 2022, the United Nations report (the UNEP branch) which measures the gap between the need for GHG emissions reductions and actual reductions ruled that "no credible pathway is in place to reach 1.5 °C."[30]

Notice that we can easily recognize the actual intention of the report's authors (UNEP) in the subtitle of their publication: *A closing window - The climate crisis requires rapid transformation of societies*.[31]

- In February 2023, the *Scientific American* reported the findings of a new report from the University of Hamburg that examined ten factors that could influence the global ability to meet Paris Agreement targets (e.g., UN governance, regulation and litigation, protests and social movements, consumption profiles, journalism, etc.). On a global scale, none of these will enable the necessary decarbonization by 2050, according to the authors. In the end, they recommend investing more in adaptation policies.[32]

- According to the World Bank, China emits 27% of the world's CO_2 and is the world's largest GHG emitter. To achieve net-zero emissions by 2060, the "Bank report estimates that it would take between US$14 and US$17 trillion of investment in green infrastructure, energy and transport technologies." The Bank believes that private sector involvement will be crucial if China is to achieve carbon neutrality...[33]

- According to Reuters, many analysts believe that the droughts of 2022 that affected China's hydroelectric capacity forced the closure of factories. This experience prompted the country to avoid over-reliance on intermittent energy sources such as solar and wind power. Since then, the construction of the equivalent of 100 coal-fired power stations has been authorized, equivalent to the energy needs of the whole of England, or two power stations a week.[34]

These are just a few of the facts that underscore the difficulty, if not the impossibility, of realizing the mythical project of the United Nations and the IPCC's coalition of nations.

But more importantly, an intrinsically physical aspect is worth mentioning CO_2 is a gas that *remains in the atmosphere for a long time.* "Between 30% and 50% of the carbon dioxide emitted today will still be present in 100 years' time; between 15% and 30% in 1,000 years' time," Koonin writes.[35] No matter how much we reduce CO_2 emissions, it will take centuries to bring the concentration down. As for the goal of stabilizing CO_2 concentration at its current value, it's just not possible *unless we eliminate all emissions everywhere on the planet right now*!

Moreover, the implementation of Net-Zero policies meets with strong resistance from the population when they have an impact on the cost of living and citizens' livelihood. Increasingly, the political class is struggling to convince the electorate of the need for such far-reaching transformations. Radical policies that are too costly and ubiquitous media harassment do not bring about the sense of urgency desired by governments. In fact, the opposite is true: the population becomes distrustful and disaffected. Add to this the instability of fossil fuel supplies from Russia, and you have a context where countries' energy security is taking precedence over the IPCC's CO_2 emissions reduction agenda. In short, realpolitik is the order of the day.

For example, on July 31, 2023, the UK government issued a press release announcing that hundreds of licenses would be granted to extract oil and gas from the North Sea. The goal: to reduce UK's energy dependency in the face of Russia's supply cut-off.[36] On July 26, 2023, Tony Blair even declared that England should not become obsessed with reducing CO_2 emissions. In his view, it would be futile to impose major sacrifices on our fellow citizens, when they could be eclipsed by other countries, such as China.[37]

Meanwhile, India generates 12.4% more electricity from fossil fuels. The world's third largest emitter of CO_2, its power plants are expected to burn 8% more coal in the fiscal year beginning April 1, 2023.[38] In Germany, the closure of three nuclear power plants makes their electricity system even more fragile and vulnerable; shortages and rising energy prices make their industries less competitive.[39] Already, there are signs of an exodus of companies and workers. Finally, protest movements are increasingly rising up to thwart the Net-Zero agenda and its experimental energy transition.

We can also see that solar and wind power deployment has shown its limitations, not least that of achieving social acceptability for such projects. On the manufacturers' side, despite substantial subsidy programs that exasperate the taxpayer, wind farm contracts have been cancelled; soaring costs are not ensuring profits.[40]

Solar energy is playing an important role in the energy transition. Here too, the sector is strongly supported by government subsidies, and growth targets are impressive. As in the case of wind power and the battery sector for the transformation of the car fleet, it is hard for citizens not to notice, among other things, the damage done to the environment when extracting certain essential raw materials. Think of the living conditions of the Congolese people when mining for lithium, of the coal mines as far as the eye can see in China and India, and of the potential graveyards of "non-renewable" waste from "renewable" energy sources such as wind and solar power.

Furthermore, the risk of depending on a country which holds a virtual monopoly on the production of key components is a cause for concern. "Today, China has more than 80% of all solar panel manufacturing stages on its territory, including polysilicon, ingots, wafers, cells and modules."[41] Many countries have suffered from relying on Russia for their fossil fuel supplies. Will we soon be dependent on China for so-called renewable energies and for batteries on which the transition to electric cars is hinging?

* * *

In short, are the ambitions of the United Nations and the IPCC relevant, economically viable and even feasible? I think not. And in any case, we're a long way from a consensus on the great Net-Zero project, with its unattainable targets and totally unrealistic timetable.

Disagreement won't be expressed openly by scientists who depend on research funding, but others dare to speak out.

When Lomborg wrote *False Alarm* in 2020, for example, the chapter entitled *Why the Paris Agreement is failing* clearly shows that he was not impressed by Al Gore's outburst calling it "a bold and historic moment." Unfortunately, he says, they were all wrong. "The Paris Agreement will cost a fortune for very few results." What's more, he calculated, to reach the 2 °C target, "every year between now and 2100, additional cuts in CO_2 equivalent to what is promised in the Paris Agreement would have to be made."[42] According to Lomborg, panic is never a good advisor; it distracts from more important problems and more effective solutions.

For Michael Shellenberger, the issue is crystal clear: as a supporter of nuclear power, he sees the damage that certain so-called renewable and green energies are doing to the environment and biodiversity. As an example, he cites the unfortunate consequences of Germany's decision to exclude nuclear power plants from its energy portfolio, making it dependent on fossil fuels from Russia. Today, the transition to a wind and solar portfolio has weakened Europe's largest economy. More recently, his organization is fighting against what he sees as a "grossly over-subsidized and inefficient wind industry" to save right whales on America's East Coast.[43] In *Apocalypse Never*, Shellenberger also denounces the hypocrisy of many so-called 'green' activists who are backed by the fossil fuel industry.

More importantly, for Shellenberger, as for Lomborg, global warming should not be solved on the backs of countries and populations in desperate need of the energy that would ensure them prosperity and a better quality of life.

As for climate risk and uncertainty specialist Judith Curry, she believes we need to remember that the underlying aim of climate policies is to improve the lot and quality of life of current and future populations. However, there is a danger that the huge investments linked to Net-Zero will be so focused on CO_2 that all other issues will be sidelined. "In the midst of this instability, she writes, we are seeing the inevitable clash between alarming proclamations about the climate crisis, the priorities of food and energy and poverty reduction, and the costs and difficulties of transitioning to net-zero CO_2 emissions."[44]

A month before COP21, where the Paris Agreement was signed, Koonin wrote an editorial in *The New York Times*. After digging into the issue for several years, he says, "he understood the science and the societal challenges well enough to see that a straightforward synthesis of a handful of basic facts led directly to the conclusion that even stabilizing human influences was so difficult as to be essentially impossible."[45]

All things considered; I think that this is exactly what we're seeing.

Thoughts...

In the months and years to come, more and more publications will reach the conclusion that the Net-Zero project can be justified neither in terms of relevance nor on the basis of a quantitative cost/benefit analysis. It will be recognized as nothing more and nothing less than an ideological initiative coupled with a very successful marketing campaign. After 30 years of magical thinking, experimentation at great cost to the taxpayer, and indirect UN governance on climate issues, governments will now have to prepare themselves for more realistic analyses and examine evidence-based alternatives.

A good place to start would be to read one of the authors of the 5th IPCC report of 2013. In *The End Is Not Yet Near*, Richard Tol explains that we are far from the apocalypse. In his opinion,

> "First, on closer inspection, the predicted impacts of climate are not nearly as bad as some would like us to believe. Climate change is a problem, for sure, and the world would be better off without it. But it is not the apocalypse.
>
> Second, the worst impacts of climate change are symptoms of underdevelopment and mismanagement. That means that we should always ask ourselves what the best way is to improve the lot of future people. Is it greenhouse gas emission reduction or economic development?" – Richard Tol.

A good question, is it not? A little reminder that this environmental economist refused to sign the IPCC report in which he participated, to protest against the alarmist tone of the conclusions.

P.S. He is, and still is, a scientist.

The *Alarmists*, the *Lovers* and the *Realists* – A recap

	CO_2-*Alarmists*	CO_2-*Lovers*	CO_2-Realists
CO_2: What is it?	A THREAT Existential threat inseparable from fossil fuels and global warming. A toxic pollutant.	PLANT FOOD Is first and foremost food for trees and plants. Helps in producing larger harvests.	A GREENHOUSE GAS Is the most significant GHG after water vapor. Is neither an existential threat nor the climate thermostat. Beneficial for plants.
Rise in the CO_2 concentration in the atmosphere Good or bad thing? Alarming or not?	IMMINENT DANGER OF A CATASTROPHE The increase is linked to fossil fuels. Alarmists are quite indifferent to the benefits of an increase in CO_2. We need to return to pre-industrial levels, they say.	A GOOD THING A good thing, as it prevents the concentration of CO_2 from falling below 150 ppm in the next glacial period. However, we produce more than we need.	NO WORRIES Current concentrations are not dangerous. More serious physiological effects on humans may begin above 2000 ppm. A level of 1000 ppm could be reached within 250 years.

	CO_2-*Alarmists*	**CO_2-*Lovers***	**CO_2-Realists**
Link between GHG emissions and the increase of CO_2 concentration in the atmosphere?	CONSENSUS The recent increase in the concentration of CO_2 in the atmosphere is largely due to the burning of fossil fuels.		
Link between the increase of CO_2 concentration in the atmosphere and Global warming	CRUCIAL IMPORTANCE The most important factor Fossil fuels ↓ CO_2 ↓ Global warming	INSIGNIFIANT EFFECT CO_2 emissions are not a significant factor in global warming. Moreover, the saturation effect will produce less and less warming for the same amount of CO_2 emitted.	HUMBLE OR MODEST EFFECT Science too immature to determine the influence of adding CO_2 to the atmosphere. Other natural influences must be explored. Saturation effect recognized.
Apocalypse and climate change catastrophism	A REAL DANGER Climate apocalypse activists.	NO If there is danger, it lies in CO_2 concentrations falling below 150 ppm.	NO Not even supported by the IPCC (WG1) Is a fabricated narrative by different interest parties.

	CO_2-***Alarmistes***	CO_2-***Lovers***	CO_2-**Realists**
Reduction in fossil fuels Required or not?	ABSOLUTELY NECESSARY Underpinning the global socio-economic and energy transformation agenda	SURE! We could cut them in half if we could find alternative energy sources. Not because they cause global warming; because they're precious and need to be conserved.	DESIRABLE BUT... Desirable but difficult. UN mitigation policies conflict with the imperatives of national governments to provide reliable and affordable energy.
Development of cleaner forms of energy. (that generate less and less CO_2) Desirable or not?	DESIRABLE BUT... Limited to solar, wind and battery energy. Don't see the advantage of switching from coal to natural gas for example. Generally against nuclear energy.	DESIRABLE BUT... Very much desirable. In fact, in a reasonable transition plan, 50% of emissions could be cut. Nuclear power is their preference. But beware of: . the risk of destabilizing the economy and energy systems with unstable energy choices. . the risk of dependence on countries with a virtual monopoly on key raw materials. . the risk of damaging biodiversity (e.g., whales on the US East Coast).	

	CO_2-*Alarmistes*	CO_2-*Lovers*	CO_2-**Realists**
Net-Zero (or Carbon neutrality) Possible or not by 2050?	IT IS A 'MUST' AND URGENTLY NEEDED It could be done but to achieve Net-Zero, major socio-economic and transformational change must happen.	CONSENSUS: NOT POSSIBLE Not possible, but there is good reason to migrate to less and less emitting CO_2 sources of energy. However, this will have to be done at a pace that is acceptable to the population and according to a plan that includes the development of the poorest countries in terms of energy. In the long term, this plan must include nuclear energy.	

CHAPTER 8

Where are we now?

By the time I began to wrap my head around climate change, it was seen as the cause of all evil on earth. On a near-daily basis, the media carried eye-catching headlines blaming extreme weather events on CO_2. The idea that global warming was *dangerous* and that *mankind was entirely responsible* had taken on the air of a dogma that could only be questioned at great risk of losing all credibility. There was a "scientific consensus," we insisted. CO_2 was raising the globe's thermostat, so all we had to do was reduce the use of carbon-dioxide-generating fossil fuels, and in so doing, mankind would make amends for having disrupted life on earth and its climate to such an extent.

For over 30 years now, the IPCC has acted as an undisputed scientific authority on the subject, and a coalition of leaders has joined the 1992 *United Nations Framework Convention on Climate Change's project*. The global plan: reduce emissions from fossil fuels, opt for Net-Zero, a transition to clean energies that do not emit carbon dioxide.

In the absence of a counterweight in the mainstream media, doubt was nevertheless expressed in a few alternative outlets. Over the past few years, others have been making their voices heard: Nobel Prize laureates, physicists, environmentalists, ecologists, economists, and other specialists in environmental public policy. They express themselves through the podcasters who receive them, through texts on their *Substack* distribution platform, for example, and through the publication of essays on the subject.

These ideas are rarely reported in the mainstream media, and the debates they could raise are consequently hushed up. The sacralization of consensus, like any sacralization of a single way of thinking, has its own perverse effects, as we can see today. Aside from the reproduction of characteristics specific to religion (a moral code, a foretold end of the world, the certainty of holding the truth, sin and the hope of redemption, etc.), we are increasingly observing in this case an exacerbated lack of confidence in the spokespersons of climate alarmism, and total confusion about the real state of our knowledge. What is *really* happening? Who can we trust?

At this point, I think I've heard a wide variety of dissenting voices, and that's fine. After all, we're talking about science, and science is a process of arguing. I confess I'm not surprised by the gap I've observed between the real state of "climate science" and what the promoters of the climate emergency are conveying. I understand better the exasperation of the people I present in this book at the media coverage and the sweeping statements made by the authorities of the United Nations and the IPCC. Exploring different opinions and research, one cannot help but suggest a change of tone and a break from allusions to the "existential threat of climate change."

Here's why:

Will you draw me a consensus?

Firstly, the idea that the warming observed since 1850 is *dangerous and essentially due to* CO_2 *emissions* is still the subject of debate among scientists. Contrary to popular belief, science is more nuanced. I found that there are three opposing views: those of the Alarmists, the Lovers, and the Realists.[3] There may be more.

As we saw earlier, when the *United Nations Framework Convention on Climate Change* was being written, there was no science to support global warming, its causes (GHGs) or its dangerous nature... Since then, the science of the IPCC has shown that there has indeed been a warming of around 1.1°C since 1850, and that the use of fossil fuels is largely responsible for the increase in CO_2 concentration in the atmosphere. However, *what science has not demonstrated* is whether there is a cause-and-effect relationship between these two phenomena. In other words, *there is no consensus that a higher concentration of CO_2 is <u>the main cause</u> of global warming.*

For example, a brand-new study published in the journal *Climate*[1] and conducted by 37 scientists from 18 countries. Its conclusion? That 40% of the warming observed in the northern hemisphere is linked to the effects of urbanization, and that the remainder can be explained by variations in solar intensity. So, little to do with CO_2... A major part of global warming would therefore be an urban problem!

As for the *dangerous nature* of global warming, despite all the IPCC's efforts to associate extreme weather events with it, the IPCC itself and many scientists see no link between them for the vast majority. What's more, there are too few reports on the benefits of global warming, such as the greening of the planet or the effect on harvests, for example.

---

[3] See table at the end of Chapter 7 for the different visions.

In this sense, the picture painted by the IPCC is incomplete and biased.

In my opinion, then, two questions remain open, the answer to the first being critical for the second:

- What is *the contribution of human activity* (CO_2) *to the overall internal and external influences* on climate warming?
- What would be *the effect of* CO_2 *emission reductions* on the climate?

An IPCC more political than scientific

Reading various authors, reports and reviews on the state of climate science, I must confess to having been surprised by the political nature of the IPCC's origins and work.

Reviewing its history and controversial episodes, we quickly realize that this is "directed science," and has been since the very beginning of the IPCC's formation. Misleading statements by political leaders, verbal puffery, literature reviews biased in favor of an official narrative, censorship and dismissal of divergent scientific opinions, pressure on scientific publications to “cancel” research: all this undermines the credibility of its leaders and the scientists who keep their disagreements quiet.

In short, after more than 30 years in office, the IPCC is increasingly being used for political ends, rather than being the guarantor of a fair, complete and rigorous picture of climate science. Let's not forget that trillions (thousands of billions) of dollars are allocated to a United Nations project that is essentially based on a consensus that is more political than scientific.

When we know that fear is the preferred tool of the politico-media class, is it really wise to trust it so blindly and to adhere to the great "transformational changes" proposed by the United Nations, the IPCC and its COPs?

An evolving science

Another reason to reject the notion of an *existential threat* and tone it down is to realize that climate science is still in its infancy.

Climate is an extraordinarily complex system, and in such a world of uncertainties and unknowns, the very notion of consensus is flawed. Real-world measuring instruments are being refined but are still the subject of much debate. Take the way we measure global temperature, for example. Even then, it's not clear-cut - just imagine!

Another discovery for me was the extent to which "climate science" is based on computer models. With its more or less reliable data, assumptions and approximations. The world of climate models and scenarios is certainly fascinating, but in the end, they are incapable of reproducing the past. In any case, even the IPCC admits that the most cataclysmic of them is outdated. Unfortunately, it is still used as a reference model in many research studies. What's more, systematically, decade after decade, it has been empirically shown that computer model predictions are overheated when compared with real-world observations. Useful for scaremongering, perhaps, but totally useless for sensible public policy.

The "science" that comes from the IPCC is the exact science that doesn't deign to have an external, independent and transparent review process. We have come to understand that it is a biased science whose mission is to validate the consensus premises of its Framework Convention. Yet this is the science that is presented to us in alarmist statements, and the one on which the world's governments are resting in their pursuit of a project that some describe as chimerical. As we have seen, if anything is impossible, it's achieving the Paris Agreement's net-zero target.

The risks of the precautionary principle

When we talk about drastically and rapidly reducing greenhouse gases (Net-Zero), we assume that this is based on empirical data resulting from observation and experience. It is also assumed that there are competent people who have assessed the risks of such social and economic upheaval by carrying out cost/benefit analyses of different scenarios. That the feasibility and consequences of such a transformational project have been properly assessed in all parts of the world.

None of this is the basis of the Net-Zero project of the United Nations and the IPCC. Rather, it is the precautionary principle, enshrined in the Framework Convention, that underpins its orientations and policies. By applying this principle, scientific certainty is no longer required; a risk of serious or irreversible damage is sufficient. As Judith Curry explains, "the rhetoric of precaution has held that we should act anyway to eliminate fossil fuels, just in case. This perspective was codified by the *United Nations Framework Convention on Climate Change* in 1992, the Kyoto Protocol in 1997 and the Paris Climate Agreement in 2015."[2]

If the precautionary principle doesn't ring a bell, it should. It was what was applied during the pandemic. Governments and public health officials may have referred to "science", but the reality was that we were in the middle of an experimental phase of messenger RNA vaccines. Today, Swedish historian Johan Norberg offers a comparative analysis of health policies in different countries. The conclusion is "uncomfortable for governments that engaged in a reckless, unprecedented pandemic experiment", he says.

> "This is a lesson for the next disaster—whatever it is, and whenever it strikes. Harsh pandemic restrictions were often defended with reference to the precautionary principle—do not take a particular course of action before an abundance of evidence is available. But there was no evidence indicating that drastic restrictions made sense.
>
> In times of uncertainty it doesn't seem like a precaution to put all your policy eggs in one basket and add to the burden of a health emergency by undermining communities, the economy, and education. Instead, it seems like negligence."[3]

After nearly 40 years of global warming alarmism, and now that we are more aware of the limits of science, the question arises: have we learned nothing? Isn't it time we paid more attention to the effects of global warming on our populations than to imagine that we can control the climate with a push-button labeled "CO_2"?

The wrong target

Another aggravating factor in our perception of the danger of a few degrees of warming is that which blames climate change for all the earth's problems, whatever they may be.

When social movements, climate activists, governments and the media carelessly identify "climate change" as the reason for all the world's problems, a normal reflex is to want to solve the problem urgently and at all costs.

But by reacting in this way, on the one hand we often miss the real target, and on the other we end up prioritizing solutions that won't solve the problem at all and, neglecting more pressing issues. I know I've already quoted Richard Tol, energy, environment and climate economist and former IPCC author, but I think if I've learned anything from my readings, it's this:

> " [...] is that many of the more dramatic impacts of climate change are really symptoms of mismanagement and poverty and can be controlled if we had better governance and more development."[4] – Richard Tol.

As Koonin explains in *Unsettled*, the residential air pollution affecting 2.4 billion people in poor countries is not the result of climate change. It's the result of poverty!

Nor are many other problems attributed to climate change. Outdated infrastructure, poorly targeted social measures and neglected water purification policies also have consequences for the environment. Attributing them to climate change merely offers an escape route for irresponsible governments all too happy with the leniency and complicity of a 4th Estate all too busy crying wolf rather than doing its homework.

The use of intimidation and censorship

There's another reason why we can afford to doubt, or at least question, the science as presented by UN officials.

If it were so indisputable, if the review process were so rigorous, why so much acrimony towards those who question and challenge it? Doesn't the scientific approach require openness to criticism, the reproduction of experiments, and transparency in the data and methodologies used?

Yet that's not the impression I get when we learn that certain studies are dismissed or that scientists who dare to add nuance to the "consensus" are labelled climate deniers. Nor is it what I get when we resort to smearing and censoring people whose opinions differ from the official narrative.

In *Climate Rationality: From Bias to Balance*, law and economics expert and environmental law professor Jason Scott Johnson asks:

> "[...] what to do about the climate advocacy industry that has produced some very interesting science, but also succeeded in moralizing scientific disagreement in a counterproductive and indeed dangerous way."[5]

For him, the 'moralization of scientific disagreement', the labeling as "deniers" of those who challenge the scientific advocacy produced by government-funded organizations, does not bode well. There comes a time when the taxpayer wonders whether it is justified to fund "science" whose purpose is not so much to evaluate the best policies, but to justify an already pre-determined policy response. In other words, is it really "science" or "directed science"?

Unfortunately, as was the case during the pandemic, it seems that once again, for some of the leading champions of alarmism and climate urgency, tolerating doubt is unbearable.

Zero-carbon transition: an inevitable retreat?

There's a lot going on in the world of Net-Zero energy transition. Several players have joined the Great Zero-Carbon Experiment and are placing their bets. New lobbyists are appearing on the market to grab the subsidy windfall, governments are betting on one technology or another, and major investment funds and central banks are taking part in "green" financing. As for the citizen, he or she is keeping his or her fingers crossed that there are people who will be able to evaluate the costs and advise when the risks are too high in relation to the insignificant gains. In the meantime, crony capitalism will receive good press and governments will be able to drape themselves in virtue.

What form of "clean" energy will emerge as the winner of the great energy transition project? It's hard to say, but there are a few pointers. As we'll see later, solar and wind power are not helping developing countries to lift themselves out of poverty, for example. They may look good in a wealthy country's diversified portfolio, but the need for social acceptability makes deployment difficult. And are they as "green" and ecological as they're made out to be?

That said, even if the major transition to energies that emit little or no CO_2 is justified and well underway, I've come to the conclusion that the outright elimination of fossil fuels in the short term is neither possible nor beneficial for humanity as a whole. A more gradual transition to increasingly clean energies is the only one that is possible, on all levels: social, economic and above all political. In fact, I'm seeing a growing retreat on the part of certain governments.

As I write these lines, the UN General Assembly is taking place in New York. True to form, Secretary-General Guterres declares that humanity's reliance on fossil fuels has "opened the gates of hell."[6] He's not happy. The G20 countries are not sufficiently reducing their dependence on fossil fuels. At the same time, however, on the other side of the Atlantic, UK Prime Minister Rishi Sunak announced a major overhaul of his government's climate policies, in what *The Guardian* describes as one of the biggest U-turns of his tenure.

> "We have reached a consensus on the future of our country that seems to satisfy no one", says Sunak.
>
> "We seem to have opted for an approach that will impose unacceptable costs on hard-working British families, costs that no one has ever really been told about and that may not be necessary to achieve the emissions reductions we need."[7]

Deadlines were postponed, policies were nuanced, and a search for alternative solutions is undertaken. Sunak is now proposing to the British "a middle road between those who want far tougher climate action and those who do not believe humans are causing climate breakdown at all, adding that the UK was already ahead of other countries in taking action."

I predict that we'll be hearing more and more speeches like this one (particularly in the run-up to elections). Because, faced with the unrealistic agenda of the United Nations, governments are confronted with a population that is rising up, challenging policies it deems insane and expressing its discontent. The fast-approaching deadlines of the Paris Agreement call for abrupt and even radical changes. Politicians are measuring the social acceptability of far-reaching transformational change... and it's becoming increasingly rare when it concretely affects citizens' wallets and lifestyles.

And then there's inflation, which is "eroding the European climate consensus," reports Bloomberg.[8] From the UK to Germany, the political class is backing down on certain green policy elements. Belgium, Holland and France are reviewing certain requirements. In short, I believe the world is entering a phase of realpolitik.

As many scientists agree, there are very good reasons for reducing countries' dependence on fossil fuels, and for switching from a highly polluting form of energy that affects air quality and important sources of drinking water, to a less polluting form of energy. However, the project must be realistic, fair, and acceptable.

The moral dilemma of zero-carbon emissions

Speaking of fair and acceptable projects, there's a theme that I haven't yet touched on in this book, but which is very present in the works of our authors. That issue is the impact of Net-Zero on the poorest countries. Is carbon neutrality to the detriment of developing countries? Are we denying them access to energies that have ensured the prosperity of our societies and the great leap towards a quality of life never before achieved in human history?

Whether it's Koonin, Curry, Shellenberger, Lomborg or Pielke, they all agree that there is something deeply immoral about imposing choices and agendas on developing societies that in no way correspond to their needs.

At the time of the 6th IPCC report, UN Secretary-General Guterres declared that all nations must join the great *CO_2 Coalition* and reduce their emissions... "out of concern for the poorest and most vulnerable nations."[9] How ironic it is to hear people who are supposedly horrified by wealth inequalities, preventing poor countries from accessing inexpensive energies that are essential to their development. And make no mistake, that's exactly what they're doing.

Shellenberger in *Apocalypse Never :*

> "Today, the World Bank is diverting funding from cheap and reliable energy sources like hydroelectricity, fossil fuels, and nuclear, too expensive and unreliable ones like solar and wind. And in October 2019, the European Investment Bank announced it would halt all financial of fossil fuels in poor nations by 2021."[10]

In their great wisdom, he recounts, the IPCC authors in 2018 proposed that poorer countries should refrain from using the energies of more prosperous countries (hydroelectricity, natural gas, nuclear power). Thus, "the leadership of Western NGOs and UN agencies, as well as many IPCC authors, have participated in efforts during the two decades to redirect public and private funds away from cheap energy and toward unreliable and expensive renewables."[11]

It begs the question, he says, "what is the real goal for these countries: sustainable development or permanent poverty?"

In Lomborg's *False Alarm*, the words used to describe the influence of the United Nations on the energy policies of the poorest countries are merciless:

> "We need to be aware that when we insist, as part of foreign aid packages, that the developing world align with our climate priorities, we are enacting a kind of imperialism. We are not listening to what the citizens of these countries want. We are jeopardizing their opportunity to lift their populations out of poverty for the sake of our own concerns. This isn't just bad policy. It's grossly unethical."[12]

In his view, the "monomaniacal focus" and the obsession with Net-Zero have resulted in the neglect of issues that genuinely improve the human condition: schools, health, nutrition, and so on.

As for Judith Curry in Climate Uncertainty and Risk, "sacrificing the well-being of today's population by restricting access to energy seems neither moral nor just."[13] She quotes a Kenyan energy expert who describes the imposition of emission reductions on poor countries as "energy apartheid"[14]. This runs counter to the real priorities of developing countries,

> "[...] the World Bank, the International Monetary Fund (IMF) and other development banks are being pressured to do the same. The African Development Bank is increasingly unable to support large natural gas projects in the face of European shareholder pressure."[15]

Reducing emissions "out of concern for the poorest and most vulnerable nations," Mr. Guterres? Prioritizing Net-Zero to the detriment of infrastructure development in the most vulnerable countries? This is not the opinion of Nigeria's Vice President, who denounces the hypocrisy of Western countries:

> "The greatest hypocrisy is that some of the largest private European and US firms are developing natural gas in Ghana, Mozambique, Nigeria, and Senegal to export to Asia and Europe, since it cannot be used in the countries of origin for lack of infrastructure."[16]

Eco-colonialism, eco-imperialism, sustainable poverty policies, green colonialism, doomsday imperialism: all terms that describe a Net-Zero policy that ignores development priorities in the "poorest and most vulnerable" countries. Conspiracy theorists? From people who deny global warming and want to discredit the work of the IPCC?

Not really... A November 2022 study analyzed 556 scenarios from the *IPCC Working Group III's 6th Report*. "Our results show [...] that the scenarios do not take into account the energy needs of southern countries to achieve their development goals."[17] In other words, "GDP per capita, energy consumption and use remain very high in developed countries, while most developing regions are expected to remain at very low levels of income, consumption of goods and services, and energy." The authors conclude on the need to revise IPCC models and scenarios in order to "make the planet 'livable' for all and not just for certain parts of the world's population."

"This is monopolistic imperialism disguised as green altruism," writes Shellenberger in a text entitled *Imperialism of the Apocalypse.*[18] Hundreds of millions of dollars are being lent to South Africa by the French and German Development Banks, for example, on the condition that they are used exclusively for renewable energies. "The Europeans hope to shift the 7.6 billion dollars currently invested by South Africa in electricity infrastructure will be redirected towards renewable energies to the detriment of coal." he reports.

In short, there is growing criticism from poorer countries that are being asked to prioritize the Net-Zero agenda over their own development. The predictable consequence? "If the West doesn't want to invest in developing the oil and gas reserves of the countries of the South, China and Russia will," warns Shellenberger. This may already be happening.

Are renewables really clean?

The enthusiasm for solar and wind power in recent decades reflects the direction set by the United Nations and the desire of investors and governments to offer an alternative to fossil fuels. The electric battery also offers hope of transforming the car fleet to reduce CO_2 emissions. You'd expect the environmental community to applaud these initiatives. But no. There's no consensus among the "greens" either. Here too, there are skeptics.

In an article entitled *The Rising Chorus of Renewable Energy Skeptics,*[19] journalist Andrew Nikiforuk offers a literature review and an overview of the concerns of the "renewable energy skeptics". The green techno-dream is so destructive that we need to think of another solution, they say. Replacing future energies with renewables would be an 'illusion', and the environmental impact would be far more damaging than that of oil extraction! You'd think they'd be lobbyists for the evil oil lobby, but they're not. They're very concerned environmentalists. Technology and the mining industry are neither "green" nor clean, in their view. "*Rare earth metals are dirty to mine and dirty to process,*" they claim. Dirty energies, as Quebec Premier François Legault put it, referring to his qualifying Alberta's "dirty oil"... but who, incidentally, is getting Quebecers to invest in a battery manufacturing plant.

In short, rare metals are anything but green, to the point where we're witnessing the most impressive *greenwashing* operation in history, according to renewables skeptics. And what about the exploitation of extremely poor populations in the Congo?

I have to admit I was shaken when I read one of the texts referred to by the journalist, *Why mining to make renewables will destroy the planet,*[20] by Alice Friedemann. The figures are staggering. But then, once again, it's hard not to get cold feet. How can we trust that our national governments have sufficiently analyzed the consequences of Net-Zero before imposing such a destabilizing and inequitable transformation on the whole of humanity?

The futility of the Kyoto protocol and of the Paris agreement

There is one very important piece of information worth highlighting at this point. I delayed a little before covering this aspect in my text. I shouldn't have. Because once we know the real impact that the Kyoto Protocol or the Paris Agreement would have had, two thoughts come to mind. Either you think you don't understand a thing, or you start to wonder about the real motivations of the United Nations, the IPCC and our governments in fooling us as they have. So I decided to dig a little deeper.

You see, the impression that we get from the work of the United Nations and the IPCC is that by reducing GHG emissions in line with the Paris Agreement targets, we will have corrected the upward trend in global temperature. However, this is not at all the case. I understood Koonin's assessment of the Paris Agreement in *Unsettled*.

> "Although some view the Paris Agreement as a crucial step toward mobilizing national and international action to mitigate global emissions, it will hardly do anything directly to reduce human influences on the climate given the 100 percent reduction needed to stabilize the carbon dioxide concentration.
>
> The modest reductions in 2030 emissions that would be achieved if all Paris pledges were met in 2030 are a long way from what would be needed to see emissions vanish."[21]

And then there was this image of a study by Bjorn Lomborg illustrating what the Paris Agreement could represent *in relation to what was needed* to maintain a warming of 2 °C. As far as I know, Bjorn Lomborg was one of the first to point this out very soon after the agreement was signed in 2015.

Look at the image and note the insignificance of the Paris Agreement, whose commitments represent only 1% of what would be needed to keep warming to 2.0 °C.

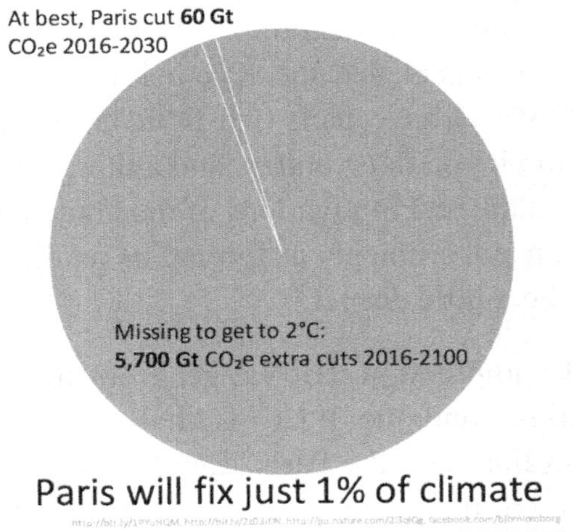

Finally, in a recent study published by the *Macdonald-Laurier Institute* entitled *Adaptation needs greater focus on climate policy*[22], economist Ross McKitrick explains quantitatively how we can conclude that the United Nations' global GHG reduction policy has failed. Using data from the *IPCC's 5th WG1 Assessment Report* and *BP's Statistical Review of World Energy*, here's what we come up with:

	Anticipated results in 2100
KYOTO PROTOCOL	CO_2 CONCENTRATION IN THE ATMOSPHERE[4] Reaching 700 ppm by 2100 is delayed by a couple of years
PARIS AGREEMENT	TEMPERATURE RISE IN 2100 Even considering an optimistic scenario, compliance with the Paris Agreement will result in a temperature reduction of 0.17°C by 2100.

Another unfortunate consequence is so-called "carbon leakage", the phenomenon whereby GHG-emitting industries move to countries with fewer regulations (China, India). Following the implementation of the Kyoto Protocol, economists demonstrated with empirical data that there was indeed "carbon leakage," enough to cancel out the reductions achieved in the countries participating in the Protocol.

I don't know about you, but this kind of thing just knocks me down... After picking myself up and realizing not only the unfeasibility of GHG reduction policies, but also their futility, I'm still holding out hope. If we can't drastically reduce GHGs, then policies of innovation, economic development and adaptation will take on their full importance.

Economic development and Adaptation policies

In a context of so many uncertainties and unknowns, we'd love to find a magic solution that would solve all the problems of pollution, poverty and the environment, while

---

[4] Here, McKitrick uses the conclusions of a study by climate scientist Tom Wigley (1998) simulating the effects of the Kyoto Protocol.

ensuring prosperity for all the world's inhabitants. We'd also like to be sure that future generations will be able to enjoy a quality of life at least equivalent to our own.

I haven't found that magical solution. Perhaps it doesn't exist. On the other hand, the authors quoted in this book suggest some interesting paths. With a little more lucidity and realism, a combination of alternatives is emerging.

An important step is to adopt a more gradual and realistic energy transition. As we have shown, Net-Zero by the end of the century is not going to happen. No matter how much Guterres cries “fire,” energy demand is bound to increase. But the good news is that there are energies that emit fewer GHGs and pollution than others. The switch from coal to natural gas is a good example of a realistic transition. Wasn't this one of the major factors behind the United States' reduction in GHG emissions since 2005? Hydroelectricity is also a clean energy source. As for nuclear power, it is said to be the cleanest form of energy available. Can it come to the rescue and meet growing needs?

The excellent article *Le grand retour du nucléaire*[23] by Hélène Baril in *La Presse* paints an interesting picture. Many countries are lining up for nuclear power. The importance of energy independence and the need to produce Zero-CO_2 energy are motivating countries to rethink their strategic energy planning. According to the *International Atomic Energy Agency*, public and private institutions are actively working to bring Small Modular Reactor (SMR) technology to fruition this decade.[24] In July 2022, the members of the European Union (EU) approved the classification of (natural) gas and nuclear power as "green" energy sources[25]. On August 10, 2023, China installed components for the first commercial SMR in its Hainan province. All excellent news, especially as China is the world's largest emitter of GHGs.

The world needs energy, and ideally, clean energy. The next step is to realize that the prosperity of populations must not be relegated to the background of a GHG reduction policy. No population will accept a lower standard of living, least of all in developing countries. Unfortunately, however, this is the effect of elevating Net-Zero to the top of the planet's priorities. It will fail, for sure. Realpolitik will force a retreat. And carbon-neutrality is not an end in itself; it's a means to an end. If other solutions can temper the negative effects of climate change, why not consider them? Especially as even the CO_2 reduction targets laid down in the Kyoto and Paris agreements have an almost insignificant impact on CO_2 concentration and warming by 2100.

Finally, we come to adaptation measures, to which Bjorn Lomborg and Judith Curry devote an entire chapter, and Steven Koonin, an important part of a section on solutions to counter the damage caused by climate change. Obviously, the richer a country is, the more it will have the means to adopt adaptation measures that will make its population and the environment less vulnerable. The more a country has access to cheap, reliable energy, the more it will be able to afford effective adaptation measures.

It will have the means to supply energy to lessen the effects of heat waves or extreme cold, or to isolate populations from the risk of flooding. Studies show that the reduction in deaths during periods of extreme heat is essentially attributable to adaptation policies (including air conditioning). We have already seen the extraordinary drop in deaths following extreme weather events. Wealthier countries have building codes that lessen the impact of severe weather. A third of the

Netherlands is below sea level! Today, the country is working with Bangladesh to mitigate the effects of flooding. Coastal protection, high wind warning systems and proper forest management all require investment but have proved effective.

Have governments put too many eggs in one GHG reduction basket? Do other choices need to be made to counter the effects of climate change rather than imagining that they can control the climate from the CO_2 control knob?

As we've seen, Net-Zero is diverting colossal sums of money into energy transition policies that will have very little effect on the upward trend in atmospheric CO_2 concentration and temperature. What should we be asking of our governments, so that they come up with solutions that are more pragmatic than ideological, more effective than sterile?

No. There is no *climate crisis*. But after 30 years of indirect governance by the United Nations and the IPCC, we need to think again. Clearly, the course we are currently on is not yielding results and does not bode well for the future.

As I said, perfect solutions do not exist. But the future of energy is taking shape before our very eyes, and the time has come to adopt solutions that are more pragmatic than ideological. Countries will continue to develop, and it's only natural that people will want to improve their quality of life while also caring for the environment and its biodiversity.

So I'll leave you with a thought-provoking paragraph from McKitrick. I think it gives a good indication of what we should be expecting from our governments from now on.

> "As the costs of mitigation policy continue to skyrocket it is incumbent on policy-makers to step back and ask some

important questions, beginning with what they are trying to accomplish. Climate change sometimes gets conflated with industrial policy.

If a policymaker's real aim is to facilitate the growth of new industries or to stimulate certain kinds of technological change then that should be stated clearly and the best options for that goal should be studied.

If the goal is to find the climate policy path that yields the maximum benefits for the lowest cost then much more attention needs to be paid to the role of adaptation, especially since, as I will show, mitigation and adaption are not necessarily complements. It may be a case of choosing one or the other."

CONCLUSION

Back in 2011, when I wrote *Pour en finir avec le Gouvernemaman* (*Enough with the Nanny State*), I didn't think it possible that states could encroach even further on the democratic values, economic freedom and primacy of human rights and liberties that characterize Western societies. Yet here we are, in a world where governments are legislating to censor the content of media platforms and where they are busy modifying economic structures that have improved the quality of life as never before in human history.

But then, I needed to account for the non-elected supranational governance entities such as the UN and the IPCC. In every age, the political leech must choose which vein to cling to. Right now, it's the climate change apocalypse.

Let's be clear. Climate change cannot be denied. The planet has a life of its own apart from that of the human species. Its climate fluctuates over the millennia, and lifeforms have no choice but to adapt. Recently (since 1850), following a cooler period known as the Little Ice Age, global temperature has risen by 1.1°C, although there have been a few decades of cooling, during which it was even thought that a new ice age was returning.

Nor can we deny that fossil fuels contribute to the increase in the atmospheric concentration of CO_2. Is this what caused global temperatures to fluctuate by one degree Celsius since the beginning of the pre-industrial era? Some doubt it, while others are so convinced that they attribute full responsibility for the recent decades' global warming to human activity. And then there are those who refuse to be so adamant.

Acknowledging the complexity of the climate system and the natural influences at work, they seek to ascertain *how much* of the warming can be attributed to the rising concentration of CO_2 in the atmosphere and vigorously denounce climate alarmism. Does this make them "global warming deniers"? Of course not, and those who mute their doubts are clearly not being helpful. I recognize they did a lot of "muting" during the pandemic, but now it's time to reconnect with what makes us human: curiosity, doubt, and reason.

Doubt is healthy. Skepticism is healthy. Questioning and adding nuance is not "denying". It is a sign of a healthy democracy. Pressing scientists and governments to provide more explanations is perfectly legitimate, as climate science is still evolving, and governments are fallible, to say the least.

What can be rejected, however, is the idea that climate change is an existential threat to the planet. That this is a matter of utmost urgency because the warming of recent decades has only had disadvantages that threaten life on Earth. But let's put our minds at rest. The climate has *always* fluctuated. We are not the first human beings to witness the climatic moods of our planet. Remember. Climate is not the weather, and according to the IPCC, there is no increase in the frequency or intensity of most extreme weather events.

As for the United Nations-orchestrated solution to "control the climate", I think it's high time we called it what it is. Net-Zero, the outright elimination of fossil fuels, is a delusion - a delusion in the sense that it makes people believe that the radical elimination of fossil fuels is a viable project and a solution to global warming. In fact, it is neither viable nor realistic for all the reasons outlined in the previous chapters. Worse still, even if nations were to respect the commitments made in the 2015 Paris Agreement, this would have no effect on the concentration of CO_2 in the atmosphere, since these commitments are insignificant compared to what it would take to achieve zero emissions by 2050. Add to this the fact that China now produces more emissions than all other countries combined. Let's face it. Net-Zero is a failure.

Climate is a compelling subject. This ball we live on is fascinating. My appreciation and understanding have grown of the people who work so hard to try to understand the many influences weighing on the climate. I really enjoyed reading the authors featured in my book. I very much appreciated their approach, humility and their great sensibility to uncertainties and risks. When you consider the complexity of natural influences on climate and the changes that have occurred over millions of years, you can't help but wonder. To what extent is man really responsible for climate change? Is there not a part that could be explained by other factors?

But if there's one thing that would never have occurred to me, it's to imagine that in all this constellation of uncertainties, there's one push-button, one, that would enable humanity in the 21st century to control global warming. A control knob called "CO_2".

"Good intentions" are not my best friends. They worry me and I distrust them like the plague, especially when they are expressed by national governments, or worse, by supranational non-elected bodies. More often than not, good intentions give rise to white elephants and monstrous bureaucracies devoid of judgment and humanity.

For almost 40 years now, the United Nations and the IPCC have had carte blanche. Consensus culture has taken over; groupthink has prevailed. We were told that climate change is an existential threat to our planet. The 4th Estate supposed guardians of democracy, freedom of press, and expression, confined themselves to reverberating the official message. But after 40 years of alarmism, and now that we're more aware of the limits of science and the lure of Net-Zero, it's time to rethink the approach.

First and foremost, there are good reasons to ensure a transition to cleaner, less GHG-generating energies. Pollution kills and harms people, the environment, and its biodiversity. However, we must opt for a realistic plan that combines prosperity for all with environmental protection.

Also, rather than betting everything on a Net-Zero, better governance, economic development of the most vulnerable countries and adaptation measures have proved their effectiveness in reducing the damage caused by extreme weather events.

Finally, we must unmask Net-Zero. Let's not think that this United Nations project is a struggle waged by a benevolent left that cares for the planet, future generations, the environment, and the most vulnerable populations... against an allegedly capitalist and polluting right. No. As we can read in the work of the IPCC itself, what is being proposed is a complete transformation of our institutions, lifestyles, and economy. Look at what we're already seeing.

The major central banks and public and private investment funds are transforming into energy transition managers, to the detriment of their primary mission of fighting inflation and ensuring the best return for their asset holders. Governments, major consulting firms and virtue lobbies are designing tomorrow's economy before our very eyes, one of cronyism and new industrial policies draped in clean and green intentions. Out with market solutions, out with competition, out with the best products at the lowest cost to the consumer.

As Jesse Kline, deputy editor of the National Post's commentary section, puts it,

> "The green economy is not a capitalist economy, where companies compete with one another to produce the best product at the lowest price and countries attract businesses and talent based on their comparative advantages; instead, it is characterized by central planning and multinational firms shopping around for whichever jurisdiction can offer the biggest corporate handouts, forcing countries to compete over who's willing to pony up the biggest bribes."[1]

How the "benevolent left" doesn't balk at this type of economy is beyond comprehension. Indeed, let's not kid ourselves. The political agenda of climate change does not lie on a left/right axis. Because if it were, the "left" would be in an uproar. This is simply pure ideology. And ideologists are people in a hurry.

To begin with, if it were coherent, it would not be so resistant to a more pragmatic transition to clean energies that nonetheless reduce GHG emissions in the short term. It would not reject out of hand solutions that would make natural gas available to the major coal-using countries, cutting their CO_2 emissions in half. The radical fringe of the net-zero emissions movement would also favor deploying of nuclear power plants and encourage research into new technologies.

It would be worried about the industrial cronyism of the big multinationals, who seduce governments, indirectly steer political choices, pick winners, and have the risks and failures absorbed by the citizen busy earning a life of his own and ensuring his family's financial security.

It would object to the developed countries' eco-colonialism policies and be concerned by the fact that its own energy transformation and the race to make batteries are largely driven by the use of coal, a major GHG emitter in countries like China and India. Speaking of batteries, it would also investigate the working conditions of workers in the gigantic mining deposits from which the materials for battery production are extracted, and the effects of this industry on the environment and biodiversity of poor countries.

No. This is no longer a left that cares about humanity's wellbeing, the environment, and all living species. We're dealing here with a new, extremely influential class of activists and upstarts, trained for 40 years in the ideology of climate catastrophism. Let's call the Net-Zero activists what they are: radicals who meet with no resistance from political and media elites who have chosen to be complicit in a display of green virtue rather than do their job. What motivates them has little to do with improving the quality of life of humans and the environment. On the contrary, such an uncompromising vision inevitably leads to an agenda of degrowth and sustainable poverty, renamed "transformational change".

Controls, regulations, "green" taxes, violation of private property, sidelining of major sectors of activity, new monitoring, and surveillance bureaucracies, even censorship and public condemnation of people expressing dissenting opinions: this is where the vision promoted by the United Nations and its IPCC is taking us: degrowth for some, sustainable poverty for others.

It's time to get a grip. Degrowth is not an option, and denying developing countries access to cheap, reliable energy is immoral.

In short, the Net-Zero project is a minefield. It's time to review everything: its *scope*, *objectives*, *expected results*, and the *conditions for its success*. More importantly, to bring it back to where democracy should be exercised: at the national level, where the ultimate judge is the electorate.

Climate alarmism cannot be used as an excuse to give the Net-Zero project such a far-reaching scope and such a brutal timetable. The single goal of eliminating all fossil fuels may be easy to sell to schoolchildren, but adults should know that things aren't that simple. Similarly, the expected results and conditions for success are magical thinking, given that national governments will inevitably be confronted with the needs and priorities of their own populations.

In short, it's time for realpolitik. Let's be a little more curious about what's in store for us, allow ourselves to doubt, to take nothing for granted, and to ask questions.

ANNEX – Tricks to guard against alarmism

Here are a few tips I've found in the works of the authors cited in this book.

1. DON'T RELY ON CATCHY TITLES – Take the time to read the article. There may be a big difference between the title and the actual content of the article. What's more, the title may simply be a quote.
2. CHECK WHETHER THE TEXT IS BASED ON PROJECTIONS OR ON OBSERVATIONAL DATA – If there are a lot of conditional verbs, that's a good clue: they're most likely projections.
3. WATCH FOR TEMPERATURE PROJECTIONS – Many studies and projections are based on the IPCC's most alarmist scenarios. Anything based on IPCC scenario 8.5 is outdated and can be disregarded.
4. CHECK WHERE THE QUOTED EXTRACT COMES FROM – There can be a big difference between the findings of the IPCC's WG1 team and their Summary Reports. Even more so with the UN Secretary-General's press releases.
5. CLIMATE IS NOT WEATHER – Remember that climate is not weather. An extreme weather event in no way indicates how the region's climate has changed in recent decades.
6. THE 97% ARGUMENT - Recourse to the 97% argument is not credible, but there's no point in trying to convince those who continue to refer to it.
7. EXTREME WEATHER EVENTS – When attributing such events to climate change and CO_2, it's important to remember that human influence can take many forms, but above all that it is very modest compared to natural influences and internal climate variability.

8. THE IMPORTANCE OF TAKING ADAPTATION INTO ACCOUNT – All too often, studies with pessimistic conclusions have failed to take into account people's ability to adapt. Find out if this is the case.

9. THE IMPORTANCE OF SOURCES AND NUMBERS – Strong statements (e.g. "Two billion people are at risk of exposure to dangerous heat") should provide a reference for the study.

10. CHECK TIME SCALES – Illustrations of upward or downward trends should specify the time scales used. The longer the time scale, the better. Otherwise, there's a risk of cherry-picking. Also, it may be a natural cycle illustrating the internal variability of certain phenomena.

11. TRY TO DETECT THE NARRATIVE – It can also be fun to try and identify the narrative from which the information is presented. Here are a few examples:
    - It is because of Climate Change
    - It's the Apocalypse
    - The identification and condemnation of individuals or groups (political or otherwise) who diverge from the official IPCC narrative. The term "denier" is used extensively in such comments.
    - Cheerleading for the side you support, without actually communicating useful information. In other words, climate activism.

About the author

Joanne Marcotte first came to public attention in 2006, when she directed the political documentary *The Quiet Illusion (L'Illusion tranquille)* in which she vigorously debunks the myth surrounding the "virtuous Quebec model".

In 2007-2008, she took part as Vice-President in the *Castonguay Committee on the financing of the Quebec healthcare system*. Then, in 2010, she co-founded the *Réseau Liberté-Québec (2010-2013)* with the intention of bringing together center-right forces in Quebec.

Five years after *L'Illusion tranquille*, Joanne Marcotte returns in 2011 with her essay *Pour en finir avec le Gouvernemaman*, in which she denounces the hyper-bureaucratization and bankruptcy of the Quebec state.

In 2023, she is back with a new essay, *Inconvenient doubts - Climate Change Apocalypse: Really?* in which she invites her readers to claim the freedom to discuss the alarmist climate change narrative's plausibility.

Joanne has been commenting and analyzing political news on her personal blog and in selected media since 2009. She holds a bachelor's degree in computer engineering from Université Laval and worked for several years in the field of computer systems architecture.

Table of Contents

NOTES

Introduction

[1] MARCOTTE, Joanne. *Le blogue de Joanne Marcotte.* https://jomarcotte.wordpress.com/

[2] MARCOTTE, Joanne. "Changements climatiques : Criminaliser le doute. Vraiment?" *Le blogue de Joanne Marcotte,* December 14, 2015. https://jomarcotte.wordpress.com/2015/12/14/changements-climatiques-criminaliser-le-doute-vraiment/

[3] MARCOTTE, Joanne. "COP21 : Le fantasme d'un 'changement de civilisation' par une économie décarbonisée." *Le blogue de Joanne Marcotte,* December 7, 2015. https://jomarcotte.wordpress.com/2015/12/07/cop21-le-fantasme-dun-changement-de-civilisation-par-une-economie-decarbonisee/

[4] ORJOLLET, Stéphane and BOTTOLLIER-DEPOIS, Amélie. Agence France-Presse. "L'inaction s'apparente à un suicide collectif." *La Presse,* November 7, 2002. https://www.lapresse.ca/actualites/environnement/2022-11-06/cop27-en-egypte/l-inaction-s-apparente-a-un-suicide-collectif-estime-le-chef-de-l-onu.php

[5] CBC. "UN Chief Says World is on a 'Highway to Climate Hell' without urgent actions to cut emissions." November 7, 2022 https://ici.radio-canada.ca/rci/en/news/1930866/un-chief-says-world-is-on-a-highway-to-climate-hell-without-urgent-action-to-cut-emissions

[6] KOONIN, Steven E. *Unsettled : What Climate Science Tells Us, What It Doesn't, and Why It Matters.* BenBella Books, 2021.

[7] LOMBORG, Bjorn. *False Alarm : How Climate Change Costs Us, Hurts The Poor and Fails To Fix The Planet.* Basic Books, 2021.

[8] SHELLENBERGER, Michael. *Apocalypse Never : Why Environmental Alarmism Hurts Us All.* Harper, June 2020.

[9] CURRY, Judith. *Climate Uncertainty and Risk : Rethinking Our Response.* Anthem Press. June 2023. Kindle Edition

Chapter 1 – Climate disinformation: sure, but which one?

[1] SHELLENBERGER, Michael. "On Behalf of Environmentalists, I Apologize For The Climate Scare." E*nvironment Progress website*, June 2020. https://environmentalprogress.org/big-news/2020/6/29/on-behalf-of-environmentalists-i-apologize-for-the-climate-scare?rq=apologize

[2] REUTERS. "One in five UK children report nightmares about climate change." March 2, 2020. https://www.reuters.com/article/climate-change-children/one-in-five-uk-children-report-nightmares-about-climate-change-idUSL1N2AV1FF

[3] SHELLENBERGER, Michael. *Apocalypse Never*, p. 27

[4] SHELLENBERGER, Michael. *Apocalypse Never*, p. 30

[5] DUMAY, Clotilde. "Notre maison brûle : quand Emmanuel Macron s'inspire de Jacques Chirac." *Le Figaro*, August 23, 2019. https://www.lefigaro.fr/politique/le-scan/notre-maison-brule-quand-emmanuel-macron-s-inspire-de-jacques-chirac-20190823

[6] SHELLENBERGER, Michael. *Apocalypse Never*, p. 30

[7] YADVINDER MALHI. "Does the Amazon provide 20% of our Oxygen?" *Ayadvinder Malhi's blog*, August 24, 2019. http://www.yadvindermalhi.org/blog/does-the-amazon-provide-20-of-our-oxygen

[8] SHELLENBERGER, Michael. "Why Everything They Say About the Amazon Including That It's the 'Lungs of the World' Is Wrong." *Forbes*, August 26, 2019. https://www.forbes.com/sites/michaelshellenberger/2019/08/26/why-everything-they-say-about-the-amazon-including-that-its-the-lungs-of-the-world-is-wrong/?sh=678048e05bde

[9] SHELLENBERGER, Michael. *Apocalypse Never*, p. 30

[10] LOMBORG, Bjorn. *The Skeptical Environmentalist – Measuring the Real State of the World.* Cambridge University Press. August 2001.

[11] THE ECONOMIST. "Doomsday Postponed." September 6, 2001. https://www.economist.com/books-and-arts/2001/09/06/doomsday-postponed

[12] LOMBORG, Bjorn. *Smart Solutions to Climate Change.* Cambridge University Press, September 2010.

[13] PERREAULT, Mathieu. "Bjorn Lomborg : un sceptique de moins." *La Presse,* September 25, 2010. https://www.lapresse.ca/dossiers/les-

grandes-entrevues/201009/25/01-4326636-bjorn-lomborg-un-sceptique-de-moins.php

[14] LOMBORG, Bjorn. *Cool It – The Skeptical Environmentalist's Guide to Global Warming*. Vintage, Kindle Edition. September 2007.

[15] TIME. "Special Report On Global Warming." April 3, 2006. https://content.time.com/time/covers/0,16641,20060403,00.html

[16] BORN, Dorothea. "Bearing Witness : Polar Bears As Icons for Climate Change Communication in National Geographic." *Environmental Communication*, February 2018. https://www.tandfonline.com/doi/full/10.1080/17524032.2018.1435557

[17] LOMBORG, Bjorn. *Cool it*, p. 4

[18] CROCKFORD, Susan. *Polar Bears Facts & Myths*. CreateSpace, 2016, p. 29

[19] CROCKFORD, Susan. "Published field study observations – not population size – prove polar bears are thriving." *Polar Bear Science Blog*, February 23, 2023. https://polarbearscience.com/2023/02/23/published-field-study-observations-not-population-size-prove-polar-bears-are-thriving/

[20] LOMBORG, Bjorn. "Partisan 'fact-checkers' Spread Climate Change Misinformation." *Washingston Street Journal*, January 26, 2023. https://www.wsj.com/articles/partisan-fact-checkers-spread-climate-misinformation-polar-bear-population-hunting-afp-temperature-11674768511

[21] Mittermeier, Cristina. "Starving Polar Bear Photographer Recalls What Went Wrong." *National Geographic*. https://www.nationalgeographic.com/magazine/article/explore-through-the-lens-starving-polar-bear-photo

[22] BLAKE, Emily. "National Geographic Says It Went 'Too Far' With Emaciated Polar Bear Video." *CBC*, August 17, 2018. https://www.cbc.ca/news/canada/north/emaciated-polar-bear-response-1.4788259

[23] KOONIN, Steven E. *Unsettled*, p. 183

[24] KOONIN, Steven E. *Unsettled*, p. 9

[25] KOONIN, Steven E. *Unsettled*, p. 188-189

[26] SHERMER, Michael and KOONIN, Steven E. "A Podcast with Michael Shermer – What Climate Science Tells Us, What It Doesn't, and Why It Matters (Steven Koonin)." November 2021. https://www.youtube.com/watch?v=Fy3EiaP4Y3Y

[27] COOK, John. "Quantifying the consensus on anthropogenic global warming in the scientific literature." *Environmental Research Letters*, May 15, 2013. https://iopscience.iop.org/article/10.1088/1748-9326/8/2/024024

[28] Gayathri Vaidyanathan. "How to Determine the Scientific Consensus on Global Warming." *Scientific American* , July 24, 2014. https://www.scientificamerican.com/article/how-to-determine-the-scientific-consensus-on-global-warming/

[29] OBAMA, Barack. *@BarackObama tweet*, May 16, 2013. https://twitter.com/BarackObama/status/335089477296988160?s=20

[30] CURRY, Judith. *Climate Uncertainty and Risk*, p. 15

[31] CURRY, Judith, Climate Uncertainty and Risk », p. 15-16

[32] CURRY, Judith. *Climate Uncertainty and Risk*, p. 19-20

[33] CURRY, Judith. *Climate Uncertainty and Risk*, p. xviii

[34] McKITRICK, Ross Dr. "The hockey stick is wrong and result of bad science", *YouTube*, July 18, 2008. https://www.youtube.com/watch?v=-1k4mFZr-gE

Chapter 2 – The science of climate change

[1] KOONIN, Steven E. *Unsettled*, p. 151

[2] Id.

[3] KOONIN, Steven E. *Unsettled*, p. 28

[4] HACHEY, Isabelle. "Il faut s'adapter à la crise climatique". *La* Presse, May 4, 2023. https://www.lapresse.ca/actualites/chroniques/2023-05-04/il-faut-s-adapter-a-la-crise-climatique.php

[5] KOONIN, Steven E. *Unsettled*, p. 26

[6] EL-HAI, Jack. "In 1975, Newsweek Predicted A New Ice Age. We're Still Living with the Consequences." *Longreads*, April 13, 2017. https://longreads.com/2017/04/13/in-1975-newsweek-predicted-a-new-ice-age-were-still-living-with-the-consequences/

[7] Id.

[8] KOONIN, Steven E. *Unsettled*, p. 38

[9] Id.

[10] KOONIN Steven E. *Unsettled*, p. 40

[11] MOORE, Patrick. "Presentation at the *Steamboat Institute,* March 2022." https://www.youtube.com/watch?v=lX1z_6pvM-Q

[12] REVUE MYSTÈRES ET SCIENCES. "L'effondrement des civilisations de l'âge de bronze." June 27, 2022. https://mysteres-et-sciences.com/2022/06/27/raisons-effondrement-civilisations-age-bronze/

[13] WIENER, James Blake & MARTIN, Caroline. "Interview: The Mysterious Bronze Age Collapse with Eric Cline. *World History Encyclopedia*, September 19, 2019.

https://www.worldhistory.org/article/1446/interview-the-mysterious-bronze-age-collapse-with/

[14] An auditor of the 6th IPCC Report. "Who Erased The Medieval Warm Period". *No tricks Zone*, September 2021.

https://notrickszone.com/2021/09/02/ipcc-6th-climate-report-who-deleted-the-medieval-warm-period-tracks-lead-to-the-university-of-bern/

[15] Id.

[16] MANN, Michael E., BRADLEY, Raymond S., HUGUES, Malcolm K. "Global Scale Temperature Patterns and Climate Forcing Over The Past Six Centuries." *NATURE*, Vol 392, April 23rd, 1998. https://www.nature.com/articles/33859

[17] IPCC AR5 WG1. "Climate Change 2013 - The Physical Science Basis." Figure 5.7, p. 409

www.ipcc.ch/site/assets/uploads/2018/02/WG1AR5_Chapter05_FINAL.pdf

[18] KOONIN, Steven E. *Unsettled*, p. 42

[19] IPCC AR6 WG1. "Climate Change 2021 – The Physical Science Base, Summary for PolicyMakers.", 2021.

https://www.ipcc.ch/report/ar6/wg1/downloads/report/IPCC_AR6_WGI_SPM_final.pdf

[20] Id., p. 10

[21] PETERSON, Jordan and KOONIN, Steven E. "A Podcast with Jordan Peterson - Unsettled: Climate and Science", *YouTube,* Rf. 36:50. January 16, 2023. https://www.youtube.com/watch?v=reaABJ5HpLk

[22] UNITED NATIONS Framework Convention on Climate Change. 1992

https://unfccc.int/process-and-meetings/what-is-the-united-nations-framework-convention-on-climate-change

[23] LEWIN, Bernie. *Searching for the Catastrophe Signal – The Origins of the Intergovernmental Panel on Climate Change*. GWPF Books, November 2017, p. 266

[24] WMO. "The Changing Atmosphere Implications for Global Security, Toronto, Canada, 27-30 June 1988." 1988 Report. https://www.cmosarchives.ca/History/ChangingAtmosphere1988f.pdf

[25] LEWIN, Bernie. *Searching for the Catastrophe Signal, p.* 269

[26] LEWIN, Bernie. *Searching for the Catastrophe Signal*, p. 265

Chapter 3 – Extreme meteorological events

[1] UNITED NATIONS. "Secretary-General Calls Latest IPCC Climate Report 'Code Red For Humanity' Stressing Irrefutable Evidence of Human Influence." *UN Press Release*, August 9, 2021. https://press.un.org/en/2021/sgsm20847.doc.htm

[2] UNITED NATIONS. "Secretary General Warns Climate Crisis Heading Into 'Unchartered Territoires of Destruction', Urges Leaders to Heed United in Science Alarming Report." *UN Press release*, September 13, 2022. https://press.un.org/en/2022/sgsm21450.doc.htm

[3] KOONIN, Steven E. *Unsettled*, p. 97

[4] SHELLENBERGER, Michael. *Apocalypse Never*, p. 15

[5] LOMBORG, Bjorn. "Bjorn Lomborg: IPCC Global Warming Report is more chill than you have read." *The Financial Post, August 12, 2021.* https://financialpost.com/opinion/opinion-ipcc-global-warming-report-is-more-chill-than-you-have-read

[6] CURRY, Judith. *Climate Uncertainty and Risk*, p. 10

[7] PIELKE, Roger Jr. "25 Years of Climate Change Research." March 28, 2019. https://rogerpielkejr.com/pielkeonclimatechange/

[8] PIELKE, Roger Jr. "How to Understand the New IPCC Report, Part 1, Scenarios." *The Honest Broker's Substack*, August 10, 2021.

https://rogerpielkejr.substack.com/p/how-to-understand-the-new-ipcc-report

[9] PIELKE, Roger Jr. "Don't Believe The Hype." *The Honest Broker's Substack*, December 19, 2022.

https://rogerpielkejr.substack.com/p/dont-believe-the-hype

[10] EM-DAT, https://www.emdat.be/database

[11] PIELKE, Roger Jr. "Pielke's Weekly Memo #12." *The Honest Broker's Substack*, Septembre 16, 2022 https://rogerpielkejr.substack.com/p/pielkes-weekly-memo-12

[12] Id.

[13] LOMBORG, Bjorn. *False Alarm*, p. 74

---

[14] Taken from *False Alarm*, p. 74. EM-DAT, 1920-2019 for average decades 1920-29, 1930-39, until 2010-19.

[15] PIELKE, Roger Jr. "Global Disasters: A Remarkable Story of Science and Policy Success." *The Honest Broker's Substack*, January 14, 2021. https://rogerpielkejr.substack.com/p/global-disasters-a-remarkable-story

[16] UNITED NATIONS. "United Nations Framework Convention on Climate Change.", Article 1.2. https://unfccc.int/resource/docs/convkp/conveng.pdf

[17] IPCC Glossary AR5. https://apps.ipcc.ch/glossary/

[18] Van Gameren, V., Weikmans, R. & Zaccai, *L'adaptation au changement climatique* (pp. 11-22). Paris: La Découverte, E. (2014). https://www.cairn.info/l-adaptation-au-changement-climatique--9782707174697-page-11.htm

[19] PIELKE, Roger Jr. "How To Understand The New IPCC Report – Part 2, Extreme Events." *The Honest Broker's Substack*, August 11, 2021. https://rogerpielkejr.substack.com/p/how-to-understand-the-new-ipcc-report-1e3

[20] PIELKE, Roger Jr. "Don't Believe the Hype." *The Honest Broker's Substack,* December 19, 2022. https://rogerpielkejr.substack.com/p/dont-believe-the-hype

[21] IPCC AR6 WG1. "Climate Change 2021 – The Physical Science Base, Summary for PolicyMakers.", paragraph A.3.1, p. 8

[22] PIELKE, Roger Jr. "What the Media Won't Tell You About... U.S. Heat Waves." *The Honest Broker's Substack*, June 15, 2022. https://rogerpielkejr.substack.com/p/what-the-media-wont-tell-you-about-9f9

[23] Id.

[24] KOONIN, Steven E. *Unsettled*, p. 1

[25] CSSR. "Fourth National Climate Assessment (NCA4)" https://science2017.globalchange.gov/

[26] KOONIN, Steven E. *Unsettled*, p. 110

[27] Id.

[28] QI ZHAO, SHILU TONG. "Global, regional, and national burden of mortality associated with non-optimal ambient temperatures from 2000 to 2019: a three-stage modeling study." *Lancet Planet Health*, Volume 5, Issue 7, July 2021. https://www.thelancet.com/journals/lanplh/article/PIIS2542-5196(21)00081-4/fulltext

[29] IPCC AR6 2021. "Synthesis Report: Climate Change 2023." https://report.ipcc.ch/ar6syr/pdf/IPCC_AR6_SYR_SPM.pdf

[30] NASA. "Greening of the Earth Mitigates Surface Warming." Novembre 2020. https://www.nasa.gov/feature/greening-of-the-earth-mitigates-surface-warming

[31] NASA EARTH OBSERVATORY. "Global Green Up Slows Warming." February 18, 2020 image https://earthobservatory.nasa.gov/images/146296/global-green-up-slows-warming

[32] PIELKE, Roger Jr. "How To Understand The New IPCC Report: Part 2, Extreme Events." *The Honest Broker's Substack*, August 11, 2021. https://rogerpielkejr.substack.com/p/how-to-understand-the-new-ipcc-report-1e3

[33] IPCC AR6 WG1. "Climate Change 2021 The Physical Science Basis", section 11.7.1.2, p. 1585

[34] Id., p. 1595

[35] KOONIN, Steven E. *Unsettled*, p. 98

[36] KOONIN, Steven E. *Unsettled*, p. 112

[37] KLOTZBACH, Philip J., BOWEN, Steven G., PIELKE, Roger Jr., et BELL, Michael. "Continental U.S. Hurricane Landfall Frequency and Associated Damage: Observations and Future Risks." *BULLETIN of the American Meteorologic Society*, Vol. 99, Issue 7. https://journals.ametsoc.org/view/journals/bams/99/7/bams-d-17-0184.1.xml

[38] UNIVERSITY OF CALGARY. Energy Education. https://energyeducation.ca/encyclopedia/Atlantic_Multidecadal_Oscillation

[39] KOONIN, Steven E. *Unsettled*, p. 89

[40] KOONIN, Steven E. *Unsettled*, p. 119

[41] IPCC AR6 WG1. "Climate Change 2021 The Physical Science Basis", p. 1588

[42] PIELKE, Roger Jr. "How to Understand The New IPCC Report: Part 2 – Extreme Events." *The Honest Broker's Substack,* August 11, 2021. https://rogerpielkejr.substack.com/p/how-to-understand-the-new-ipcc-report-1e3

[43] Id.

[44] PIELKE, Roger Jr. "Misinformation in the IPCC." *The Honest Broker's Substack, Marsh 29,* 2023 https://rogerpielkejr.substack.com/p/misinformation-in-the-ipcc

---

[45] BRIAND, Mylène. "In pictures: Southern British Columbia flooded." Updated November 17, 2021. *RADIO-CANADA British Columbia*, https://ici.radio-canada.ca/nouvelle/1840277/inondation-fraser-merritt-abbotsford-agassiz-lillooet-vancouver

[46] RADIO-CANADA Colombie-Britannique. "Fortes pluies en C.-B. : éboulements, inondations et des milliers d'évacués.", 15 novembre 2021.

https://ici.radio-canada.ca/nouvelle/1839857/inondations-pluie-abbotsford

[47] FRASER BASIN COUNCIL. "1894 Fraser River Flood" https://www.arcgis.com/apps/MapSeries/index.html?appid=ba733ac76582486982c6464dd262237b

[48] IPCC AR6 WG1. "Climate Change 2021 – The Physical Science Base.", p. 1569

[49] KOONIN, Steven E. *Unsettled*, p. 138

[50] IPCC, AR6 WG1, "Climate Change 2021 – The Physical Science Base.", p. 1560

[51] Id, p. 1155

[52] Id.

[53] AGENCE France-PRESSE. "Aux États-Unis, accord pour préserver le fleuve Colorado de la sécheresse." *Radio-Canada Info*, 22 mai 2023.

https://ici.radio-canada.ca/nouvelle/1981391/etats-unis-accord-fleuve-colorado-secheresse

[54] KOONIN, Steven E. *Unsettled*, p. 140

[55] LÉVEILLÉ, Jean-Thomas. "Inondations en Colombie-Britannique – Quand la nature reprend ses droits." *La Presse,* November 20, 2021.

https://plus.lapresse.ca/screens/0199b8e6-73af-49ba-b912-7d2bb7949904%7C_0.html

[56] RÉMILLARD, David. "Inondations: Baie St-Paul battue de vitesse. " *ICI QUÉBEC,* 19 mai 2023. https://ici.radio-canada.ca/nouvelle/1979259/baie-saint-paul-mesures-protection-financement-inondations

[57] JUNEAU, Philippe. "Inondations dans Charlevoix – Qui l'eût cru(e)? " *La Presse,* 12 mai 2023.

https://www.lapresse.ca/debats/opinions/2023-05-12/inondations-dans-charlevoix/qui-l-eut-cru-e.php

[58] PIELKE, Roger Jr. "What the Media Won't Tell You About Wildfires." *The Honest Broker's Substack,* June 8, 2023.

https://rogerpielkejr.substack.com/p/what-the-media-wont-tell-you-about-783

[59] ALEEM, Zeeshan. "Canada's wildfire crisis could be a preview of the future." *MSNBC*, June 7, 2023. https://www.msnbc.com/opinion/msnbc-opinion/wildfire-canada-smoke-air-quality-us-rcna88184

[60] DUFOUR, Thomas. "La petite histoire des grands feux du Saguenay-Lac-St-Jean." *Le Quotidien,* 3 juillet 2020. https://www.lequotidien.com/2020/07/03/la-petite-histoire-des-grands-feux-du-saguenay-lac-saint-jean-b63c7fe1b08ad02f885050e64bb2bc0f/

[61] PIELKE, Roger Jr. "What the Media Won't Tell You About Wildfires." *The Honest Broker's Substack*, June 8, 2023. https://rogerpielkejr.substack.com/p/what-the-media-wont-tell-you-about-783

[62] SHELLENBERGER, Michael. *Apocalypse Never*, p. 5

[63] Marlon et al., 2008 – Quote taken from *False Alarm, p. 67*

[64] KOONIN, Steven E. *Unsettled*, p. 142

[65] NASA EARTH OBSERVATORY. "Building a Long-Term Record of Fire." https://earthobservatory.nasa.gov/images/145421/building-a-long-term-record-of-fire

[66] KOONIN, Steven E. *Unsettled,* p. 142

[67] LEBLANC, Étienne. "Le déni climatique gagne du terrain." *Radio-Canada Info,* June 8, 2023. https://ici.radio-canada.ca/nouvelle/1986031/changement-climat-phenomene-feux

[68] @IweatherON, June 8, 2023 tweet, 10:54 AM.

[69] National (Canadian) Forest Database. http://nfdp.ccfm.org/fr/data/fires.php

[70] CHAVARDÈS, Raphaël D. et al. "Converging and diverging burn rates in North American boreal forests from the Little Ice Age to the present." *International Journal of Wildland Fire 31(12)* , November 8, 2022. https://doi.org/10.1071/WF22090

[71] Danneyrolles, Victor, CHAVARDÈS, Raphaël, BERGERON, Yves, "Les forêts boréales nord-américaines brûlent beaucoup, mais moins qu'il y a 150 ans." *Métro*, 11 juin 2023. https://journalmetro.com/environnement/3097707/les-forets-boreales-nord-americaines-brulent-beaucoup-mais-moins-quil-y-a-150-ans/

[72] NATIONAL GEOGRAPHIC, Education Blog, "How Climate Change is Affecting World Heritage Sites." *National Geographic,* July 6,2016.

https://blog.education.nationalgeographic.org/2016/06/07/how-climate-change-is-affecting-world-heritage-sites/

[73] KOONIN, Steven E. *Unsettled,* p. 151

[74] CURRY, Judith. *Climate Uncertainty and Risk*, p. 90

[75] KOONIN, Steven E. *Unsettled,* p. 155

[76] KOONIN, Steven E. *Unsettled,* p. 157

[77] KOONIN, Steven E. *Unsettled,* p. 166

[78] CURRY, Judith. *Climate Uncertainty and Risk*, p. 130

[79] NASA, GLOBAL CLIMATE CHANGE, "Fire and Ice: Why Volcanic Activity is Not Melting the Polar ice Sheets." May 6, 2020. https://climate.nasa.gov/explore/ask-nasa-climate/2982/fire-and-ice-why-volcanic-activity-is-not-melting-the-polar-ice-sheets/

Chapter 4 – The IPCC, its climate models and scenarios

[1] Koonin, Steven E. *Unsettled*, p. 80

[2] PETERSON, Jordan and KOONIN, Steven E. "A Podcast with Jordan Peterson - Unsettled: Climate and Science", *YouTube,* January 16, 2023. https://www.youtube.com/watch?v=reaABJ5HpLk

[3] Koonin, Steven E. *Unsettled*, p. 86

[4] Id., p 87

[5] Id., p. 88

[6] Id., p. 93

[7] Id., p. 93

[8] Id,, p. 89

[9] Curry, Judith. *Climate Uncertainty and Risk*, p. 65

[10] RENEE CHO. "What Uncertainties Remain in Climate Science." *Columbia Climate School*, 12 janvier 2023. https://news.climate.columbia.edu/2023/01/12/what-uncertainties-remain-in-climate-science/

[11] HACHEY, Isabelle. "Nous n'avons encore rien vu" *La Presse*, July 15, 2023. https://www.lapresse.ca/actualites/chroniques/2023-07-15/nous-n-avons-encore-rien-vu.php

[12] Koonin, Steven E. *Unsettled*, p. 57

[13] CURRY, Judith. *Climate Uncertainty and Risk*, p. 81

[14] Koonin, Steven E. *Unsettled*, p. 56

[15] PIELKE, Roger Jr. "How to Understand the New IPCC Report: Part 1 Scenarios" August 10, 2022.

https://rogerpielkejr.substack.com/p/how-to-understand-the-new-ipcc-report

[16] Id.

[17] CLINTEL. "The Frozen Climate Views of the IPCC." May 2023, pp. 76-91. https://clintel.org/thorough-analysis-by-clintel-shows-serious-errors-in-latest-ipcc-report/

[18] PIELKE, Roger Jr., RITCHIE, Justin. "Distorting the view of our climate future: The misuse and abuse of climate pathways and scenarios." *Science Direct*, October 20, 2020. https://www.sciencedirect.com/science/article/abs/pii/S2214629620304655

[19] CLINTEL. "The Frozen Climate Views of the IPCC. " p. 7

[20] IEA - International Energy Agency. "Coal 2022 Executive Summary." https://www.iea.org/reports/coal-2022/executive-summary

[21] CURRY, Judith. *Climate Uncertainty and Risk*, p. 84

[22] Id.

[23] STONE, Madeleine. "Five climate scenarios to expect, from the most optimistic to the strangest." *National Geographic,* January 5, 2022. https://www.nationalgeographic.fr/environnement/cinq-scenarios-climatiques-auxquels-sattendre-du-plus-optimiste-au-plus-etrange

[24] NOAA. "Carbon dioxide now more than 50% higher than pre-industrial levels." June 3, 2022. https://www.noaa.gov/news-release/carbon-dioxide-now-more-than-50-higher-than-pre-industrial-levels

[25] CLINTEL. "The Frozen Climate Views of the IPCC", p. 96

[26] IPCC AR6 WG1, p. 11

[27] CLINTEL. "The Frozen Climate Views of the IPCC", p. 95

[28] MCKITRICK, Ross. "Once again IPCC's math doesn't check out." *National Post*, October 13, 2022. https://www.fraserinstitute.org/article/once-again-ipccs-math-doesnt-check-out

[29] LEWIS, Nicholas. "Objectively combining climate sensitivity evidence." *Springer Link*, September 19, 2022. https://link.springer.com/article/10.1007/s00382-022-06468-x

[30] VOOSEN, Paul. "Use of 'too hot' climate models exaggerates impact of global warming." *Science*, May 4, 2022 https://www.science.org/content/article/use-too-hot-climate-models-exaggerates-impacts-global-warming

---

[31] CLINTEL. "The Frozen Climate Views of the IPCC." May 2023, pp. 76-91 https://clintel.org/thorough-analysis-by-clintel-shows-serious-errors-in-latest-ipcc-report/

[32] CERES-SCIENCE. "New Study Suggests Global Warming Could Be Mostly An Urban Problem." September 1, 2023. https://www.ceres-science.com/post/new-study-suggests-global-warming-could-be-mostly-an-urban-problem

[33] CURRY, Judith. *Climate Uncertainty and Risk*, p. 104

[34] MANN, Michael et al. "Multidecadal Climate Oscillations During the Past Millenium Driven by Volcanic Forcings." *Science*, March 5, 2021. https://www.science.org/doi/10.1126/science.abc5810

[35] PENN STATE. "Apparent Atlantic Warming Cycle likely an Artefact of Climate Forcing." *Science News*, March 4, 2021. https://www.sciencedaily.com/releases/2021/03/210304161117.htm

[36] CURRY, Judith. "Canceling the AMO." *Climate Etc., Judith Miller's Blog,* March 6, 2021. https://judithcurry.com/2021/03/06/canceling-the-amo/

Chapter 5 – No, there is no climate crisis

[1] CURRY, Judith. *Climate Uncertainty and Risk*, p. 6

[2] GELOSO, Vincent. "La croissance économique a le pouce vert." *IEDM*, July 17, 2023 https://www.iedm.org/fr/la-croissance-economique-a-le-pouce-vert/

[3] PIELKE, Roger Jr. "Think of the Implications of Publishing." *The Honest Broker's Substack*, July 17, 2023. https://rogerpielkejr.substack.com/p/think-of-the-implications-of-publishing

[4] ALIMONTI, Gianluca et al. "RETRACTED ARTICLE : A critical assessment of extreme events trends in times of global warming."*Springer Link*, January 13, 2022. https://link.springer.com/article/10.1140/epjp/s13360-021-02243-9

[5] JOURNET, Paul. "La maladie de la croissance" *La Presse*, February 12, 2023. https://www.lapresse.ca/contexte/2023-02-12/crise-ecologique/la-maladie-de-la-croissance.php

[6] JOURNET, Paul. "Les dollars ne sont pas verts" *La Presse*, April 22, 2023. https://www.lapresse.ca/actualites/chroniques/2023-04-22/jour-de-la-terre/les-dollars-ne-sont-pas-verts.php

[7] GHEBREYESUS,Tedros. "Climate Change Is Already Killing Us." *Foreign Affairs*, September 23, 2019.

---

https://www.foreignaffairs.com/articles/world/2019-09-23/climate-change-already-killing-us

[8] OMS, « Air pollution », https://www.who.int/health-topics/air-pollution#tab=tab_2

[9] KOONIN, Steven E. *Unsettled*, p.171

[10] MURPHY, Rex. "Elon Musk and Twitter are the least of CBC's worries." *The National Post*, 23 avril 2023. https://nationalpost.com/opinion/musk-and-twitter-are-the-least-of-cbcs-worries

[11] LEGENDRE, Benjamin, PAYSANT, Delphine. "Place à l'ère de l'ébullition mondiale, dit l'ONU." *Le Devoir*, July 27, 2023.

https://www.theguardian.com/world/video/2023/jul/27/the-era-of-global-boiling-has-arrived-warns-the-un-video

[12] SCHIFFMAN, Richard. "Les jeunes générations peuvent-elles surmonter leur éco-anxiété." *National Geographic*, July 2, 2022. https://www.nationalgeographic.fr/environnement/les-jeunes-generations-peuvent-elles-surmonter-leur-eco-anxiete

[13] LEBEL, Anouk. "Un dépliant distribué à Outremont propose de faire un enfant de moins pour protéger l'environnement." Journal de Québec, May 19 2023. https://www.journaldequebec.com/2023/05/19/un-depliant-distribue-a-outremont-propose-de-faire-un-enfant-de-moins-pour-proteger-lenvironnement

[14] LEGAULT, Jean-Benoit. "Les jeunes Canadiens seraient plus écoanxieux que jamais." *Le Devoir*, March 20, 2023. https://www.ledevoir.com/societe/785967/environnement-les-jeunes-canadiens-seraient-plus-ecoanxieux-que-jamais

[15] GOV. Of PAKISTAN Finance Division. "Pakistan: Flood Impact Assessment" *Economic Survey 2010-2011* https://www.finance.gov.pk/survey/chapter_11/Special%20Section_2.pdf

[16] PLAUTZ, Jason. "The Environmental Burden of Generation Z." *Washington Post Magazine*, February 3rd, 2020. https://www.washingtonpost.com/magazine/2020/02/03/eco-anxiety-is-overwhelming-kids-wheres-line-between-education-alarmism/

[17] HEC Research and Consulting. "*Rapport soumis à l'Agence de santé publique du Canada. Ce que nous avons entendu : perspectives sur les changements climatiques et la santé publique au Canada*." April 23, 2023. https://www.canada.ca/fr/sante-publique/organisation/publications/rapports-etat-sante-publique-

canada-administrateur-chef-sante-publique/etat-sante-publique-canada-2022/ce-que-nous-avons-entendu-perspectives-changements-climatiques.html

[18] KROL, Ariane. "Le CISSS de Laval s'attaque à son empreinte carbone" *La Presse*, April 17, 2023. https://www.lapresse.ca/actualites/environnement/2023-04-17/planete-bleue-idees-vertes/le-cisss-de-laval-s-attaque-a-son-empreinte-carbone.php

[19] Id. 13

[20] PEARCE, Fred. "It's not just Climate : Are We Ignoring Other Causes of Disasters?" *Yale Environment 360*, February 8, 2022. https://e360.yale.edu/features/its-not-just-climate-are-we-ignoring-other-causes-of-disasters

[21] LOMBORG, Bjorn. *False Alarm*, p. 221-222

[22] McKITRICK, Ross. "The flaw in relying on worst-case-scenario climate model." *Financial Post*, June 23, 2020. https://financialpost.com/opinion/ross-mckitrick-the-flaw-in-relying-on-a-worst-case-scenario-climate-model

[23] McKITRICK, Ross. "Junk Science Week : Ross McKitrick: The Social Cost of Carbon game." *Financial Post*, May 25, 2023. https://financialpost.com/opinion/junk-science-week-social-cost-of-carbon-game

[24] CURRY, Judith. *Climate Uncertainty and Risk*, p. 20

[25] UNITED NATIONS. "*World Population Prospects 2022*", p. 27 https://www.un.org/development/desa/pd/sites/www.un.org.development.desa.pd/files/wpp2022_summary_of_results.pdf

[26] HUMAN PROGRESS. https://www.humanprogress.org/why-are-we-so-gloomy/

[27] KOONIN, Steven E. *Unsettled*, p. 173

[28] EURONEWS. "Wolves, bears and bison : 50 species make 'spectacular' comeback in Europe." September 27, 2022. https://www.euronews.com/green/2022/09/27/wolves-bears-and-bison-50-species-make-spectacular-comeback-in-europe

[29] IPSOS, "Environnement : le climato-scepticisme progresse chez les Français." December 8, 2022 https://www.ipsos.com/fr-fr/obscop-2022-fr

[30] LEWIN, Bernie. *Searching for the Catastrophe Signal – The Origins of the Intergovernmental Panel on Climate Change*. GWPF Books, November 21, 2017. p. 148

Chapter 6 – Science or politics?

[1] LEWIN, Bernie. *Searching for the Catastrophe Signal – The Origins of the Intergovernmental Panel on Climate Change*. GWPF 2017, p. 13

[2] AARHUS Center for Science Studies. "The Study of Critical Environmental Problems (SCEP)." https://css.au.dk/en/projects/shaping-cultures-of-prediction/playground-webexhibit/disciplinary-topics/climate-modelling/scep

[3] LEWIN, Bernie. *Searching for the Catastrophe Signa,l* p. 153. https://mitpress.mit.edu/author/study-of-mans-impact-on-climate-smic-18160/

[4] Id., p. 185

[5] Id., p. 6

[6] Id., p. 211

[7] GREEN, Kenneth P. "Is Climate Catastrophe Really Ten Years Away?" *Fraser Institute*, January 13, 2022. https://www.fraserinstitute.org/studies/is-climate-catastrophe-really-10-years-away

[8] JOHNSTON, Jason S. "The Hand of Government in the Intergovernmental Panel on Climate Change." *Fraser Institute,* September 13, 2022. https://www.fraserinstitute.org/studies/hand-of-government-in-the-intergovernmental-panel-on-climate-change

[9] Id., p. 1

[10] IPCC. "Structure of the IPCC" https://www.ipcc.ch/about/structure/

[11] CURRY, Judith. *Climate Uncertainty and Risk,* p. 44

[12] JOHNSTON, Jason S. "The hand of Government in the Intergovernmental Panel on Climate Change", p. 15

[13] KOONIN, Steven E. *Unsettled*, p. 199

[14] SEITZ, Frederick. "Major deception on global warming." *Wall Street Journal*, 12 juin 1996. https://stephenschneider.stanford.edu/Publications/PDF_Papers/WSJ_June12.pdf

[15] PIELKE, Roger Jr. "A Tip From An Insider." *The Honest Broker's Substack,* May 9, 2023. https://rogerpielkejr.substack.com/p/a-tip-from-an-ipcc-insider

[16] CURRY, Judith. "UN's climate panic is more politics than science." *Climate Etc., Judith Miller's Blog*, March 28, 2023.

https://judithcurry.com/2023/03/28/uns-climate-panic-is-more-politics-than-science/

[17] JOHNSTON, Jason S. "The Hand of Government in the Intergovernmental Panel on Climate Change.", p. 19

[18] PIELKE, Roger Jr. "The Political Agenda of the IPCC." *The Honest Broker's Substack,* 15 mai 2023. https://rogerpielkejr.substack.com/p/the-political-agenda-of-the-ipcc

[19] Id.

[20] LIDSKOG, Rolf et SUNDQVIST, Göran. "Lost in transformation: The Paris Agreement, the IPCC and the quest for national transformative change.", *Frontiers in Climate,* September 20, 2022 https://www.frontiersin.org/articles/10.3389/fclim.2022.906054/full

[21] IPCC. "Profile of Jim Skea." https://www.ipcc.ch/people/jim-skea/

[22] IPCC. "IPCC elects Jim Skea as the new Chair." 26 juillet 2023 https://www.ipcc.ch/2023/07/26/jim-skea-new-ipcc-chair/

[23] Hallier, Boris. FranceInfo. "C'est maintenant ou jamais : ces défis qui attendent Jim Skea, le nouveau président du GIEC." *Franceinfo,* July 26, 2023. https://www.francetvinfo.fr/monde/environnement/crise-climatique/c-est-maintenant-ou-jamais-ces-defis-qui-attendent-jim-skea-le-nouveau-president-du-giec_5972942.html

[24] SKEA, Jim, LIEBREICH, Michael. "Ep36: Prof. Jim Skea 'It's climate, Jim, but not as we know it'.", 25 mars 2021. https://www.youtube.com/watch?v=oAWUdL5ZKsk

Chapter 7 – CO_2 and the Net-Zero craze

[1] KOONIN, Steven E. *Unsettled,* p. 50

[2] BizNewsTV. "Meet Dr Patrick Moore: Greenpeace co-founder who left the organisation hijacked by political left." *Biz NewsTV, Youtube,* 22:00 https://www.youtube.com/watch?v=GLkiQ0qlm-M&t=1985s

[3] @wideawake_media. *Twitter*, July 6, 2023. https://twitter.com/wideawake_media/status/1676934421461581824?s=20

[4] MOORE, Patrick. *Video on Twitter*, April 25, 2023. https://twitter.com/wideawake_media/status/1650791638594093059?s=43&t=46R3luO8vdKgPB6J2h1Srg

[5] WILLIAMSON, Chris and MOORE, Michael. "Greenpeace's Ex-President: Is Climate Change Fake?." *Chris Williamson's Podcast,* September 18, 2021. https://www.youtube.com/watch?v=E5K5i5Wv7jQ

[6] HAPPER, William et LINDZEN, Richard, "Challenging Net-Zero with Science.", February 2023. https://co2coalition.org/publications/challenging-net-zero-with-science/

[7] KOONIN, Steven E. *Unsettled,* p. 62

[8] PETERSON, Jordan and LINDZEN, Richard. "A Podcast with Jordan Peterson - Climate Science: What Does it Say?" *YouTube* at 50:17. https://www.youtube.com/watch?v=7LVSrTZDopM

[9] RBC Special Report. "The 2 Trillion Transition: Canada's Road to Net-Zero", October 20, 2021, https://thoughtleadership.rbc.com/the-2-trillion-transition/

[10] McKITRICK, Ross. "The Important Climate Study You Didn't Hear Abouté" *The Financial Post*, April 12, 2023. https://www.fraserinstitute.org/article/the-important-climate-study-you-didnt-hear-about

[11] UNITED NATIONS 2015. "Paris Agreement" https://unfccc.int/sites/default/files/english_paris_agreement.pdf

[12] UNITED NATIONS Climate Action. https://www.un.org/en/climatechange/net-zero-coalition

13 IEA. "Net Zero by 2050 – A Roadmap for the Global Energy Sector." May 2021. https://www.iea.org/reports/net-zero-by-2050

[14] NORDHAUS, WD. "Strategies for the Control of Carbon Dioxide." *Yale University EliScholar,* 1977. https://elischolar.library.yale.edu/cgi/viewcontent.cgi?article=1674&context=cowles-discussion-paper-series

[15] Carbon Brief Staff. "Two degrees: The history of climate change's speed limit." December 8, 2014. https://www.carbonbrief.org/two-degrees-the-history-of-climate-changes-speed-limit/

[16] Jaeger, C.C., Jaeger, J. "Three views of two degrees." *Reg Environ Change* 11 (Suppl 1), 15–26 (2011). https://doi.org/10.1007/s10113-010-0190-9

[17] POSTDAM Institute for Climate Impact Research Press Release. "The Father of the 2 Degrees Limit."October 19, 2017 https://www.pik-potsdam.de/en/news/latest-news/201cthe-father-of-the-2-degrees-limit201d-schellnhuber-receives-blue-planet-prize

[18] UNFCCC. "Fact sheet: The need for mitigation." *United Nations Fact Sheet,* November 2009. https://unfccc.int/files/press/backgrounders/application/pdf/press_factsh_mitigation.pdf

[19] IPCC. "Global Warming of 1.5." *IPCC Special Report, 2018*. https://public.wmo.int/fr/ressources/bulletin/rapport-sp%C3%A9cial-du-giec-sur-le-r%C3%A9chauffement-plan%C3%A9taire-de-15-%C2%B0c

[20] MURPHY, Robert P., McKITRICK, Ross. "Off Target: The Economic Literature Does Not Support the 1.5° Climate Ceiling." *Fraser Institute*, July 27, 2021. https://www.fraserinstitute.org/studies/off-target-the-economics-literature-does-not-support-the-1.5c-climate-ceiling

[21] Id., p. 19

[22] Id., p. 19

[23] COINTE, Beatrice, GUILLEMOT, Hélène. "A history of the 1.5° Ctarget." *WIREs Climate Change*, January 18, 2023. https://wires.onlinelibrary.wiley.com/doi/abs/10.1002/wcc.824

[24] PIELKE, Roger Jr. "The 1.5 Degree Temperature is a Dead Man Walking." *The Honest Broker's Substack,* January 23, 2023. https://rogerpielkejr.substack.com/p/the-15-degree-temperature-target

[25] Van BEEK, Lisette et al. "Navigating the political: A. analysis of political calibration of integrated assessment modelling in light of the 1.5C goal." *Science Direct. Environmental Science & Policy*. Volume 133, pages 193-202. July 2022 https://www.sciencedirect.com/science/article/pii/S1462901122001174

[26] S&P. "Global energy demand to grow 47% by 2050 with oil still top source: US EIA." October 6, 2021. https://www.spglobal.com/commodityinsights/en/market-insights/latest-news/oil/100621-global-energy-demand-to-grow-47-by-2050-with-oil-still-top-source-us-eia

[27] Id.

[28] L'Express. "Pétrole: pourquoi Shell ne réduira plus sa production d'ici 2020." June 15, 2023. https://www.lexpress.fr/economie/petrole-pourquoi-shell-ne-reduira-plus-sa-production-dici-2030-5OQAQYTJCFFS7EOJ2J3W4BFYEE/

[29] THE ECONOMIST. "The world is missing its lofty climate targets. Time for some realism." *The Economist,* November 3, 2023. https://www.economist.com/leaders/2022/11/03/the-world-is-missing-its-lofty-climate-targets-time-for-some-realism

[30] UNEP. "Emissions Gap Report 2022." October 27, 2022. https://www.unep.org/fr/resources/rapport-2022-sur-lecart-entre-les-besoins-et-les-perspectives-en-matiere-de-reduction-des

[31] UNEP. "The Closing Window – Climate crisis calls for rapid transformation of societies." 2022. https://wedocs.unep.org/bitstream/handle/20.500.11822/40932/EGR2022_ESEN.pdf?sequence=8

[32] HARVEY, Chelsea. "War, Politics, Business Make Meeting 1.5 Degrees C Target Unlikely." *Scientific American*, originally from *E&E News*, February 2, 2023. https://www.scientificamerican.com/article/war-politics-business-make-meeting-1-5-degrees-c-target-unlikely/

[33] WORLD BANK. "China's Transition to a Low-Carbon Economy and Climate Resilience Needs Shifts in Resources and Technologies." October 12, 2022. https://www.worldbank.org/en/news/press-release/2022/10/12/china-s-transition-to-a-low-carbon-economy-and-climate-resilience-needs-shifts-in-resources-and-technologies

[34] STANWAY, David & XU, Muyu. "Analysis: China's new coal plants set to become a costly second fiddle to renewables." March 22, 2023. Reuters https://www.reuters.com/business/energy/chinas-new-coal-plants-set-become-costly-second-fiddle-renewables-2023-03-22/

[35] KOONIN, Steven E. *Unsettled, p. 69*

[36] RAVIKUMAR, Sachin. "Londres va accorder des licences pour extraire du pétrole et du gaz en mer du Nord." *Reuters*, July 31, 2023. https://www.reuters.com/article/grande-bretagne-energie-idFRKBN2ZB0HA

[37] GROVES, Jason. "Tony Blair warns Britain can't stop climate change alone." *Daily Mail.com,* July 26, 2023. https://www.dailymail.co.uk/news/article-12346355/Tony-Blair-warns-Britain-stop-climate-change-Ex-PM-suggests-costly-green-policies-futile-UKs-emissions-dwarfed-Chinas-makes-break-Labour-policy.html

[38] CHYE, Matthew & CHEW, Carman. "India's power output grows at fastest pace in 33 years, fuelled by coal." *Reuters,* April 5, 2023. https://www.reuters.com/business/energy/indias-power-output-grows-fastest-pace-33-years-fuelled-by-coal-2023-04-05/

[39] BUNYAN, Rachael. "Germany faces electricity shortages and can expect to see industries leave the country due to green energy policy 'disaster' that saw nuclear power plants ditched for renewables, business chiefs warm." *Daily Mail.com,* June 7, 2023. https://www.dailymail.co.uk/news/article-12168981/Germany-faces-electricity-shortages-expect-industries-leave-country.html?mc_cid=ed6d43445e&mc_eid=6fe6aaedd4

[40] MATHIS, Will & BLOOMBERG. "As renewable-energy demand soars amid extreme heat, rising costs are making offshore wind projects so

expensive that 'it doesn't make sense to continue'." *Fortune,* July 22, 2023. https://fortune.com/2023/07/22/offshore-wind-projects-nixed-due-to-costs-despite-clean-energy-needs/

[41] NORMAND, François. "L'industrie solaire est trop concentrée en chine, selon l'AIE." *Les Affaires,* August 10, 2022. https://www.lesaffaires.com/secteurs/manufacturier/l-industrie-solaire-est-trop-concentree-en-chine-selon-laie/635087

[42] LOMBORG, Bjorn. *False Alarm*, p. 120

[43] SHELLENBERGER, Michael. "New documentary 'proves' building offshore wind farms does kill whales." *The New York Post*, August 26, 2023. https://nypost.com/2023/08/26/new-documentary-proves-that-offshore-windfarms-kill-whales/

[44] CURRY, Judith. *Climate Uncertainty and Risk*, p. 249

[45] KOONIN, Steven E. *Unsettled,* p. 212

Chapter 8 – Where are we now?

[1] CERES-SCIENCE. "New Study Suggests Global Warming Could Be Mostly An Urban Problem." September 1, 2023. https://www.ceres-science.com/post/new-study-suggests-global-warming-could-be-mostly-an-urban-problem

[2] CURRY, Judith. *Climate Uncertainty and Risk,* p. 151

[3] NORBERG, Jonah, "Sweden during the Pandemic." *CATO Policy Analysis*, Numéro 959, August 29, 2023. https://www.cato.org/policy-analysis/sweden-during-pandemic

[4] CURRY, Judith. *Climate Uncertainty and Risk,* p. 20

[5] JOHNSTON, Jason S. *Climate Rationality : From Bias to Balance.* Cambridge University Press, 2021, p. 968

[6] AHMED, Issam. "L'humanité aux portes de l'enfer, avertit le chef de l'ONU." *Agence France-Presse*, September 20, 2023. https://www.lapresse.ca/actualites/environnement/2023-09-20/crise-climatique/l-humanite-aux-portes-de-l-enfer-avertit-le-chef-de-l-onu.php

[7] CRERAR, Pippa & HARVEY, Fiona. "Rishi Sunak confirms rollback of key green targets." *The Guardian*, September 20, 2023. https://www.theguardian.com/environment/2023/sep/20/rishi-sunak-confirms-rollback-of-key-green-targets

[8] MILLIGAN, Ellen, RUDGARD, Olivia, SORGE, Petra. "The Inflation Crisis Is Fraying Europe's Climate Consensus." *Bloomberg,* September 10,

2023. https://www.bloomberg.com/news/articles/2023-09-10/inflation-crisis-is-fraying-europe-s-climate-consensus

[9] UN. "Secretary-General's statement on the IPCC Working Group 1 Report on the Physical Science Basis of the Sixth Assessment." August 9, 2021. https://www.un.org/sg/en/content/secretary-generals-statement-the-ipcc-working-group-1-report-the-physical-science-basis-of-the-sixth-assessment

[10] SHELLENBERGER, Michael. *Apocalypse Never,* p. 225

[11] Id., p. 229

[12] LOMBORG, Bjorn. *False Alarm*, p. 17

[13] CURRY, Judith. *Climate Uncertainty and Risk*, p. 251

[14] CURRY, Judith. *Climate Uncertainty and Risk*, p. 214

[15] Id.

[16] Id.

[17] Tejal Kanitkar, Akhil Mythri, T. Jayaraman. "Equity Assessment of Global Mitigation Pathways in the IPCC Sixth Assessment Report." *OSFPreprints*. November 2022. https://osf.io/p46ty/

[18] SHELLENBERGER Michael. "Imperialism of the Apocalypse." *PUBLIC's Substack,* November 11, 2022. https://public.substack.com/p/imperialism-of-the-apocalypse

[19] NIKIFORUK, Andrew. "The Rising Chorus of Renewable Energy Skeptics." *The Thee,* April 7, 2023. https://thetyee.ca/Analysis/2023/04/07/Rising-Chorus-Renewable-Energy-Skeptics/

[20] FRIEDEMANN, Alice. "Why mining to make renewables will destroy the planet." *Peak Everything, Overshoot, & Collapse blog, August 26, 2022.* https://energyskeptic.com/2022/why-how-mining-metals-for-renewables-will-destroy-the-planet-not-prevent-the-energy-crisis/

[21] KOONIN, Steven E. *Unsettled*, p. 221-222

[22] MCKITRICK, Ross, "Adaptation needs greater focus in climate policy." *Macdonald-Laurier Institute Publication,* June 2023 https://macdonaldlaurier.ca/we-need-climate-adaptation-over-climate-mitigation/

[23] BARIL, Hélène. "Le grand retour du nucléaire. ", *La Presse*, October 22, 2022. https://www.lapresse.ca/affaires/2022-10-10/transition-energetique/le-grand-retour-du-nucleaire.php

[24] IAEA, International Atomic Energy Agency. "What are Small Modular Reactors (SMRs)? ", September 13, 2023. https://www.iaea.org/newscenter/news/what-are-small-modular-reactors-smrs

---

[25] RADIO-CANADA INFO, "L'UE approuve la classification du gaz et du nucléaire comme énergies "vertes", From Agence France-Presse, July 6, 2022.

Conclusion

[1] KLINE, Jesse. "The case for doing (almost) nothing about climate change.", *The National Post,* September 28, 2023
https://nationalpost.com/feature/the-case-for-doing-almost-nothing-about-climate-change

Made in the USA
Monee, IL
04 November 2026

67383974R00134